An Assessment of
DEFENSE NUCLEAR AGENCY FUNCTIONS

Pathways To a New Nuclear Infrastructure for the Nation

NATIONAL DEFENSE RESEARCH INSTITUTE

Prepared for the Office of the Secretary of Defense

RAND

David C. Gompert, Vice President,
Director, NDRI

Acquisition and Technology Policy Center
Eugene C. Gritton, Director

Study Director
C. Bryan Gabbard

Principal Author
Richard O. Hundley

Future Environments	**Organizational Options**	**Synthesis and Analysis**
Richard F. Mesic*	John A. Friel*	Elwyn D. Harris*
Maurice Eisenstein	Dave J. Dreyfus	Bruno W. Augenstein
Scott A. Harris	Daniel H. Jones	Dave J. Dreyfus
Roger C. Molander	Wayne G. Walker	Maurice Eisenstein
Peter Wilson		Ken A. Solomon

Historical and Legal Affairs	**Special Issues Panel**	**Congressional Liaison**
William R. Harris	Robert E. LeLevier*	Scott A. Harris
	Philip E. Coyle	
	Victor Gilinsky	
	Harold M. Agnew	
	Paul J. Bracken	
	John C. Toomay	

Senior Advisory Group

Brent Scowcroft
Albert D. Wheelon
Lew Allen
Herbert F. York

Corporate Review

Paul K. Davis
Corporate Research Manager,
Defense Planning and
Technology

David S. C. Chu
Kenneth P. Horn
Bernard D. Rostker

*Task Leader.

PREFACE

This report documents the results of a congressionally directed study (expressed in Public Law 103-139, November 1993) to evaluate alternative ways of accomplishing the functions currently performed by the Defense Nuclear Agency (DNA). The motivation for the request was a desire on the part of the Congress to substantially reduce the size and costs of programs associated with the Cold War era and to resolve two different congressional perspectives about meeting the nation's needs in areas associated with DNA. The result was a four-month RAND study sponsored by the Assistant to the Secretary of Defense for Atomic Energy (ATSD/AE). It was accomplished in the Acquisition and Technology Policy Center of RAND's National Defense Research Institute, a federally funded research and development center sponsored by the Office of the Secretary of Defense, the Joint Staff, and the defense agencies.

CRITICAL FUNCTIONS

As the U.S. nuclear infrastructure is substantially reduced in size, the capabilities for three basic continuing functions must be assured:

- Maintaining a safe, secure, and reliable stockpile of nuclear weapons (and improving it, if necessary, with safety-related changes, for example).

- Understanding the effects of nuclear weapons used by or against U.S. forces.

- Supporting effective and ambitious arms-control and disarmament measures to dispose of excess nuclear weapons and material.

THE CURRENT NUCLEAR INFRASTRUCTURE

The U.S. nuclear weapons infrastructure is split between the Department of Defense (DoD) and the Department of Energy (DoE), and is further fragmented within both. Budgets and numbers of qualified staff are declining across the board, as they should, but there is a distinct danger that the decline will be managed poorly, with loss of critical capabilities. The current highly decentralized structure is poorly suited to a tightly managed large-scale drawdown.

THE DEFENSE NUCLEAR AGENCY

DNA is only one small part of the overall nuclear infrastructure. It continues to focus largely on the critical functions above, but has become increasingly active in a variety of nonnuclear weapon technology areas. A proposal has been made to transform DNA into a "special weapons agency."

ASSESSMENT OF CONGRESSIONAL OPTIONS

We assessed a number of options for accomplishing the functions now performed by DNA. Our findings were as follows:

- No other *single* organization in DoD or DoE could take on all the functions now performed by DNA, without substantial risks in some functional areas.

- However, by appropriately distributing DNA's functions among the military services, the Advanced Research Projects Agency (ARPA), DoE, and the On-Site Inspection Agency, the risks would be "manageable" in all functional areas—i.e., effective risk-mitigating measures would be straightforward—and in some areas would be only minimal.

- There are several potential pitfalls in breaking up DNA's functions: It would increase fragmentation, it would place critical nuclear tasks within organizations with high priorities for other, nonnuclear activities (perhaps leading to inattention), and it could prejudice, if not impede, potentially important steps to reorganize the larger U.S. nuclear infrastructure for the new era.

- Although dismantling DNA is feasible, doing so would likely save only $10 million to $20 million per year, because DNA releases 90 percent of its funds to other organizations, leaving relatively little overhead to be reduced by reorganization.

- Even without restructuring, DNA has instituted procedures to reduce manpower by 250 positions and overhead expenses over the next five years and thus will be evolving into a "leaner DNA." These reductions are equivalent to an annual savings of at least $20 million once they are fully implemented.

- Having DNA take on greater nonnuclear functions (as a special weapons agency, for example) would run the risk of inattention to nuclear responsibilities and competencies, but this risk could be mitigated by strong and steady management. Further, adding such functions may be necessary to retain and attract high-quality staff able also to do the nuclear functions.

OPTIONS ADDRESSING THE LARGER ISSUES OF CONSOLIDATION AND STABILIZATION

We concluded that issues regarding the overall nuclear infrastructure were far more important than the DNA issue itself, but we could examine the larger issues only briefly in this study. We concluded:

- It is questionable whether there remains any reason to continue the separation of nuclear responsibilities between DoD and DoE.

- Merger under DoE would be unwise if not infeasible, given DoD's operational nuclear responsibilities and direct concerns with nuclear effects and threat reduction. By contrast, merger under DoD merits a thorough examination of feasibility.

- Consolidation of current DoD nuclear-weapons-related support activities (stockpile support, nuclear effects research, and threat reduction and arms control) would be a good first step in this direction, and useful in its own right.

- Dismantling DNA in the meantime could make these larger solutions, which might have very large cost saving consequences, more difficult. The marginal cost of leaving DNA intact for now is negligible.

RECOMMENDATIONS

Our principal recommendations are three:

- The United States should decide how it wants to consolidate and stabilize the overall nuclear infrastructure first (and should also consider incorporating selected nonnuclear activities into that infrastructure), and *then* decide what to do with DNA.

- In the next 1–2 years, DoD should tighten its management of nuclear matters by consolidating its nuclear-weapons-related support activities under one senior federal executive in the Office of the Secretary of Defense (OSD), and within only one or a few reporting agencies.

- Consolidation within the DoD of *all* U.S. nuclear-weapons-related activities should be *seriously* considered as a primary organizational option for a more compact, sustainable, and responsive U.S. nuclear infrastructure for the 21st century. That could be studied and implemented, if appropriate, in 2–4 years.

CONTENTS

FIGURES

TABLES

SUMMARY

BACKGROUND

This report responds to a congressional request for an independent examination of options to accomplish the functions now performed by the Defense Nuclear Agency (DNA). The request was motivated by a desire to resolve two quite different perspectives about the future of DNA. One perspective would assign DNA's current functions to other organizations—the military, the Advanced Research Projects Agency (ARPA), or the Department of Energy (DoE). The other would have DNA continue as an entity, but with an expanded role in dealing with all weapons of mass destruction (WMD), and certain technical aspects of counterproliferation and arms control. Both perspectives reflected a desire to reduce the cost and size of the nuclear weapon infrastructure, but there were major disagreements in Congress about how to proceed. Against this background, the Congress requested a study examining a range of options, some of which it specified. The result was a four-month RAND study and the current report, which evaluates the congressional options and suggests others for consideration. Our analysis focused primarily on assessing the options for their effect on the government's ability to maintain critical expertise in nuclear and certain conventional defense areas, cost savings, and (to a much lesser extent) issues of schedule and implementation.

APPROACH

The nuclear weapon infrastructure developed during the Cold War includes a very large and unique set of skills and facilities. However, the relevant government organizations and industry are dispersed bureaucratically and geographically. Today, the nation must find ways to reduce this infrastructure substantially without risking a minimum essential capacity to meet its future requirements involving nuclear weapons. DNA is only a small part of this complex infrastructure and the question quite rightly raised about DNA's future should be asked, for the same reasons, about this infrastructure as a whole. Thus, it became clear early in the problem-formulation phase of the study that an evaluation of alternatives for accomplishing DNA's functions should be set not only in the emerging international security environment, but also in the context of the *overall* nuclear infrastructure. It also proved important to consider the potential value and risks of having DNA take on new nonnuclear missions as proposed by the Department of Defense (DoD) in 1993. In what follows, consistent with the structure of the full report, we summarize relevant aspects of the international security context, DNA's current and proposed functions, our analysis of congressional options for accomplishing DNA's functions, issues involving the *overall* nuclear infrastructure, alternatives for consolidating that infrastructure, and our recommendations.

THE FUTURE ENVIRONMENT AS A CONTEXT FOR ANALYSIS

Nowhere has the end of the Cold War had a more profound effect on national security policy than in the area of nuclear weapons. Dramatic changes are under way in the source and nature of the nuclear threat to the United States, the role of nuclear weapons in national strategy, the size of the U.S. nuclear arsenal, the nation's future needs for nuclear weapons and delivery systems, and the infrastructure required to meet those needs.

Predicting the future course of international events in the post-Cold War era is fraught with uncertainties. However, nuclear weapons will clearly not go away and, although they have become much less important in political, strategic, and military calculations than at the height of the Cold War, it appears that they will have a significant, albeit different, role in the years ahead. The size of nuclear arsenals is declining, and the threat posed to the U.S. homeland has been sharply reduced. However, none of the nations that currently possess nuclear weapons are likely to give them up, and other nations are attempting to acquire them. Some of these new nuclear actors may be less stable and less predictably deterred than the nuclear powers were during the Cold War. Some are hostile to the United States, and indeed may be motivated to acquire nuclear weapons in part to reduce their disadvantages in a confrontation with the United States. It is not difficult to envision futures in which the probability of nuclear weapon use is higher than during the Cold War. Nor is it difficult to imagine a weakened but revanchist Russia becoming heavily dependent on its nuclear arsenal or an ambitious China determined to challenge the substantial advantage in nuclear weapons now held by Russia and the United States. As peculiar as this may seem today, it is even possible that the United States will once again contemplate relying on its nuclear weapons to help extend deterrence to some of its allies and other key nations.

We examined a range of future environments, from a "benign" extreme of a few hundred nuclear weapons under the control of a few responsible states to the "menacing" extreme of arsenals with thousands of weapons worldwide and widespread proliferation to irresponsible actors. In view of the uncertainties regarding the future, it is prudent not to assume that the benign extreme will prevail, even though U.S. arms control and nonproliferation policies will continue to work in that direction. The United States cannot gamble on achieving the world it seeks where only a few countries have nuclear weapons and proliferation has been checked. Because nuclear weapons have unique political and military leverage, the United States must be prepared for a world in which at least some existing and future nuclear powers perceive an advantage in improving, expanding, and brandishing their nuclear arsenals. In this uncertain post-Cold War era, the United States must avoid at all costs a future in which it is at a nuclear disadvantage relative to any other country. Indeed, if the price is low enough, there may well be value in maintaining current U.S. nuclear advantages, however uncertain the future benefit or applicability of such a posture might currently seem.

This means that the United States will have a continuing though smaller need for nuclear weapons that are operationally effective and politically credible, as well as current knowledge of the effects of the use of its own and others' nuclear weapons. These concerns translate into a minimum of three important continuing requirements:

1. Diligent stewardship of the nuclear stockpile, to include safety, security, and reliability of U.S. nuclear weapons, and possible modernization needs (many of them safety-related);

2. Capacity to understand and deal with the use of nuclear weapons (and other WMD), to include accidental or intentional detonations and their effects on U.S. interests, allies, and forces;

3. Vigorous pursuit of opportunities to reduce the threat of nuclear weapons and other WMD use by reducing the number of weapons that exist and the number of nations that have them.

Meeting these requirements will give the United States the ability to face the likely futures and the means to pursue the most benign future. It is in the context of meeting these requirements that we have assessed the future course of DNA.

A PROFILE OF DNA

DNA and its predecessors were established as a source of expertise on the effects of nuclear weapons, the survivability of military systems in the presence of nuclear effects, and the related issue of weapon lethality, testing, and support for military operations planning. It was specifically established as a source of expertise for the Office of the Secretary of Defense (OSD) that would be independent of the military services and the operational nuclear commanders-in-chief (CINCs). As such, DNA was meant to play both an oversight role and audit function in the areas of nuclear system survivability performed by the services. In earlier years, DNA's predecessors were also fully responsible for all nuclear weapons in the U.S. stockpile once custody was given to the DoD. When arsenals increased to many thousands of weapons, the services were given this responsibility and DNA continued in specialized roles: accounting for, planning the movement of, and providing independent audits of the safety of the U.S. stockpile weapons in the DoD.

Today's DNA is a complex organization in transition from its near-totally nuclear Cold War role to a mixed role in the new era. After an in-depth review of its activities and budget, we clustered DNA's exceedingly diverse activities in four categories. The first three correspond to the three enduring requirements noted above. The fourth category covers various nonnuclear activities where DNA's nuclear and other competencies can be applied, and for which a national-level focal point agency would arguably be useful. The four categories are:

1. Nuclear weapons stockpile support,

2. Nuclear weapons effects (NWE) research and operational planning support,

3. Nuclear threat reduction and arms control, and

4. Conventional defense technologies. [1]

To discharge these functions, DNA oversaw the expenditure of $907 million in FY93. Of this, $540 million was spent on line items for OSD, $250 million was in cooperative threat reduction (CTR) funding directed to DNA, and $117 million was transferred to DNA from other agencies. These figures include 6.2 research, development, test, and evaluation (RDT&E) funding of $399 million (reduced to $235 million in FY94), which is a commonly used measure of DNA's budget. Roughly 90 percent of DNA's funding was provided to other organizations rather than spent in-house.

Nuclear Weapons Stockpile Support

DNA's work in "operational" stockpile support includes the development and implementation of procedures, systems, and supporting technologies to assure safety, security, reliability, and—to a limited extent—survivability of the nuclear stockpile. It conducts training

---

[1] Conventional defense technologies, as used herein, includes nonnuclear WMD technologies and related activities.

programs and independent field inspections of the services' nuclear weapons operational procedures and storage facilities. It maintains an up-to-date database on the status of all DoD nuclear storage facilities worldwide and all nuclear weapons in DoD custody and also maintains a command center and response teams to react to nuclear accidents or incidents anywhere in the world.

DNA conducts these efforts largely using in-house military personnel. In FY93 (the latest year for which complete financial data are available), expenditure for these activities was about $10 million (about $20 million if military personnel costs are included), a level that has been relatively stable over recent years and is unlikely to change dramatically in the future.

Nuclear Weapons Effects Research and Operational Planning Support

Historically, DNA has been the DoD focal point and "center of excellence" for understanding and documenting the effects of nuclear weapons on systems, i.e., for vulnerability/hardening and targeting support to the services and CINCs. It has developed hardening technologies and design approaches that mitigate the effects of nuclear weapons. DNA has also developed and operated test facilities to validate the nuclear hardness of U.S. and allied systems and to assess the vulnerabilities of adversaries' systems.

DNA funds other organizations to perform most of this function, relying on the contractor community and other government laboratories. That is, DNA serves mostly as an executive agent for activities, rather than as a direct provider. In FY93, DNA allocated $413 million to this category of work, an amount comparable to that of recent prior years. With the end of the Cold War, the suspension of underground nuclear testing, the reduction of the number of nuclear weapons in the stockpile, and the changes in U.S. military strategy and operational planning, the level of activity in this category is being sharply reduced.

At the same time, a debate has begun regarding the future need for NWE research. Does the United States already know most (or all) of what it needs to know about nuclear weapons effects to meet the needs of the post-Cold War era? Or are there questions requiring continued investigation (e.g., understanding how new systems will operate in a nuclear weapons environment)? From our prior knowledge and interviews with experts, we conclude that the United States *does* know most of what it needs to know about nuclear weapons environments to meet future requirements. This said, most of the expense of this category of work involves not the search for new physics knowledge, but complex testing in above ground nonnuclear simulators for radiation hardness of systems. Technology changes rapidly and the effect of nuclear environments on new devices or systems is not the same as it was on older technologies, even though the radiation environments may be the same. This is perhaps even more important than before as the United States comes to depend more on commercial systems for information and intelligence dominance. Without continued testing and expertise, the United States could encounter important surprises (and be unable to cope with them well) if nuclear weapons were used in a future contingency. This is by no means an endorsement of all the specific research DNA is doing, nor of the level of planned funding. It is, instead, a caution against believing that because nuclear weapons phenomena are broadly understood, no further NWE work is needed.

Nuclear Threat Reduction and Arms Control

DNA's role in arms control has grown with the increasing number and scope of treaties in recent years. DNA conducts research and development (R&D) activities supporting treaty monitoring of underground nuclear testing, strategic and intermediate-range nuclear force reductions, chemical weapons elimination, and conventional force reductions. Its current activities range from acting as an executive agent for implementing support agreements between the United States and states of the former Soviet Union (FSU) and procuring items for the safe and secure dismantling of FSU nuclear weapons within the context of the Nunn-Lugar Cooperative Threat Reduction legislation, to developing and managing an on-site inspection technology testbed. It is clear that this general area of nuclear threat reduction and arms control will play an increasingly important and prominent role in the years ahead—not only in implementing existing arms-control agreements, but in progressing toward benign futures and away from more threatening ones.

In FY93, DNA spent about $337 million in this category of work, with the vast bulk of this going for cooperative threat reduction (CTR) activities. Expenditures will probably increase significantly. DNA has only a small in-house staff dedicated to these activities. Most of the work is contracted to private companies.

Except for some activities supporting treaty monitoring (a small percentage of the effort in this category), the activities in this category involve mostly coordination and procurement functions, not RDT&E. Although they require knowledge and management competence concerning nuclear weapons, these activities have little in common with DNA's historical specialized technical functions. They are instead a major part of the reorientation of DNA currently under way.

Conventional Defense Technologies

DNA's charter allows it to "perform technical analyses, studies, and research on nonnuclear matters of critical importance to the Department of Defense where DNA has unique capabilities developed as part of its nuclear responsibilities." Activities in these "nonnuclear matters" have steadily increased as a fraction of the DNA budget over the past few years and now cover a range of selected conventional weapons areas. Within the past year, DNA has also been directed by the Assistant to the Secretary of Defense for Atomic Energy (ATSD/AE) to assume the lead in developing counterproliferation acquisition plans; it could play a larger role in this area in the future. This technical synergism between nuclear core competencies and some aspects of conventional munitions applications, when combined with similar overlaps in WMD applications areas, has provided the basis for recommendations to OSD that DNA become the Defense Special Weapons Agency (DSWA).

DNA relies on the contractor community and other government laboratories for the conduct of most of these activities. RAND's analysis of the DNA programs showed that DNA allocated about $147 million to this area in FY93. This is currently a growth area for DNA, and the expenditure numbers for FY94 and later years are expected to be considerably higher.

THE CONGRESSIONAL OPTIONS

Congress asked RAND to assess five specific options for accomplishing the functions currently performed by DNA: (1) transferring DNA's functions to the individual armed services or ARPA; (2) maintaining DNA as a separate agency, evolving to meet the conditions of the new national security environment; (3) transferring DNA's functions to the DoE nuclear weapon laboratories; (4) combining any of these previously listed options; and (5) reorganizing DNA to significantly reduce the agency's operating, management, administrative, and other overhead costs. The options were to be evaluated by assessing the effect on the federal government's ability to maintain critical expertise in nuclear and conventional defense areas and to effectively perform functions now done by DNA, the annual cost saving, if any, and a reasonable implementation schedule.

With respect to the five congressionally specified options, we conclude:

- No other *single* organization could accomplish all the functions now performed by DNA (Options 1 and 3), without substantial risks in some functional areas.

- However, if the DNA functions are distributed among the military services, ARPA, DoE, and the On-Site Inspection Agency (OSIA) in an optimum fashion (Option 4), the risks are "manageable"[2] in all areas, and in some cases only minimal. However, such a step could aggravate what we consider a major existing problem in an era of declining resources, changing requirements, and uncertain futures: the fragmentation of the nuclear weapons infrastructure within and between DoE and DoD.

- Permitting or encouraging DNA to take on greater nonnuclear weapons functions (a feature of Option 2) would run the risk of inattention to nuclear responsibilities and competencies, but this risk could be mitigated by a strong and steady management commitment to avoid such diversion. Further, adding the

---

[2]By "manageable" we mean that they can be effectively mitigated in a straightforward fashion.

nonnuclear functions may be critical in retaining and attracting high-quality staff able to do the nuclear functions.

- A "leaner" DNA (Option 5) seems feasible. The key would be to encourage management to pursue vigorously its plan to increase efficiency in the administration and direction of contracted research and other procurement activities.

Significantly, however, if major cost saving is the objective, transferring DNA's functions is not the answer. Our conclusions on cost savings are as follows:

- At most, transferring DNA's functions might provide a yearly overhead savings between $10 million and $20 million, which must be weighted against the risks associated with the various transfer options.

- DNA has instituted procedures to reduce manpower and overhead expenses over the next five years and thus will be evolving into a "leaner DNA." These reductions are equivalent to an annual savings of at least $20 million once they are fully implemented.

None of these options addressed the broader questions that face the nation regarding the nuclear infrastructure. Like DNA, other parts of DoD and DoE play important roles in meeting the requirements we believe are key for the future. Yet our concern about fragmentation pertains to this entire infrastructure. In the absence of a sound architecture for the new era of reduced resources and changing needs, actions to alter the functions now performed by DNA will not necessarily advance the national interest in the nuclear weapons infrastructure as a whole. Indeed:

- The least-risk congressional option we can identify—dispersing the functions of DNA to several entities as summarized above—could prejudice, if not impede, potentially important options involving a larger consolidation of DoD and DoE functions. We discuss that in what follows.

THE NUCLEAR INFRASTRUCTURE

Concerns About Trends in the Overall Nuclear Infrastructure

The nuclear infrastructure is dispersed across a large number of organizations in DoD and DoE. DNA is only one piece in this mosaic, accounting for less than 10 percent of nuclear infrastructure spending. Much larger are the DoE nuclear weapon R&D and production activities and the stockpile-stewardship initiatives scattered throughout DoE and DoD. Further, DNA performs functions that are vitally linked to various elements of the DoE and the DoD nuclear infrastructure, including some DoE laboratories, DoE production facilities, the services, some defense agencies, the Joint Staff, and the nuclear CINCs. This interdependent infrastructure underscores the importance of considering DNA functions within the framework of related DoD and DoE programs and the likely future evolution of both.

Indeed, although we did not analyze other organizations with nuclear weapons responsibilities in the same depth we did DNA, our research revealed widespread indications that

> **The most important question to be addressed is how to transform the nation's nuclear weapon infrastructure into one that would be much less expensive, more compact, sustainable, and able to meet the critical functions in a range of possible future contexts.**

As things stand, each separate organization is rethinking priorities and budgets. As a result of normal organizational and individual responses to the downsizing, nuclear-weapon-related activities are shifting from being uniformly high priority to relatively low priority at many if not all of these components. This is especially true of those organizations in which nuclear-related matters are not a central responsibility. But such a tendency is also observable in institutions (e.g., DNA and DoE) where nuclear issues are a central responsibility. In any case, fragmentation is occurring and many of the pieces could fall below critical mass (in terms of the quantity and quality of necessary expertise and resources).

Thus, although the overall reduction of the nuclear infrastructure is wholly justified, we conclude that the *manner* in which the individual components are downsizing is worrisome. As the management literature demonstrates, during downsizing specialized capabilities can readily be lost altogether. However, the United States cannot afford to lose nuclear core competencies in the understanding, maintenance, testing, and effects assessment of nuclear weapons.

Upon reviewing the situation, then, we are concerned that a fragmented nuclear infrastructure, even if acceptable for the Cold War, will not answer the needs of the new era for a number of reasons. First, across-the-board resource reductions raise the risk of some or all of the fragments losing effectiveness as noted above. Second, separating control of budget and resources from executive planning responsibility could erode focus and collaboration at the very time they are most needed. Third, one of the most demanding new requirements of the future—disposing of U.S. and FSU stockpiles—may demand a highly consolidated approach. And, fourth, having such a consolidated approach would be very useful as an example to others as the United States urges other nations possessing nuclear weapons to maintain strong and secure, centralized, civilian control.

Consolidation Options

An obvious option to consider, then, is consolidation to bring together the essential people and functions for maintaining critical intellectual mass and assuring management attention and effectiveness: i.e., consolidation (and streamlining) of the various nuclear activities now distributed between and within DoE and DoD. We make the case for this in broad and indicative terms influenced by our research; actual *adoption* of a plan to move in this direction would require much more extensive data and analysis.

We do believe that serious attention should be paid to creating a new management entity in DoD. This entity would carry out not only the essential nuclear functions of DNA, but also those of the Defense

Programs organization of the DoE, and the Office of the Assistant to the Secretary of Defense for Atomic Energy, which includes the On-Site Inspection Agency as well as DNA. The functions of the DoE weapons laboratories with continued involvement in nuclear weapons research would be included in this new organization.[3]

Such a consolidation could also include responsibilities involving other WMD, special purpose conventional weapons and counter-proliferation programs. In our judgment, the case for doing so would be even stronger in the event of a consolidation of nuclear functions in DoD than it is for the current DNA, since the risk of insufficient attention to the nuclear function would be less (the nonnuclear activities would be a smaller fraction of the total).

RECOMMENDATIONS

Most of the study was an effort to assess the options identified by the Congress. As discussed above, however, we concluded that a broader perspective was needed. Although we were unable to develop and analyze the option in depth (something that would require substantially more data, time, and level of effort), we were able to explore issues in enough depth to justify the following overarching recommendations:

> **The United States should decide how it wants to consolidate and stabilize the overall nuclear infrastructure first (and also consider incorporating selected nonnuclear activities into that infrastructure), and then decide what to do with DNA.**

> **In the near term, DoD should tighten its management of nuclear matters by consolidating all of its current nuclear-weapons-related support activities (stockpile support, nuclear effects research, and threat reduction and arms control) under one senior federal executive in OSD, and within one or at most a small number of agencies reporting to that executive.**

---

[3]As others have noted, it is not clear that the United States has a long-term need for all three of the current DoE weapons laboratories (Los Alamos, Lawrence Livermore, and Sandia). A "slimming down" of the weapon-related components of this laboratory structure may be in order, with a possible consolidation in one or two locations. This would in turn facilitate their merger within DoD.

Over the longer term, consolidation within the DoD of all U.S. nuclear-weapons-related activities should be seriously considered as a primary organizational option for a much smaller, but enduring and robust, U.S. nuclear infrastructure for the 21st century.

Steps should be initiated in FY95 to validate and refine concepts for a consolidation of all nuclear weapon activities under the Secretary of Defense beginning in FY96. This should include a review of all DoE nuclear-weapons-related programs and a characterization of the management, operational and technical functions, and linkages of those programs. This framework must be clear before any further steps can be considered regarding a broader consolidation of the nuclear infrastructure. The National Security Council should request DoD and DoE to develop a program and budget description that would allow serious assessment of consolidation and scale-down options in FY96. Figure S.1 shows our recommended schedule.

RAND *MR442-S.1*

STEPS	FY94	FY95	FY96	FY97

• Re-engineer DNA
  – Focused/enhanced agency
  – Support infrastructure consolidation
    planning				

• Realign DoD nuclear infrastructure
  – Single authority for nuclear (and WMD)
    planning and budget decisions				

• Review DoE national security programs
  – Nuclear activities
  – Defense activities				

• Evaluate/validate options to consolidate all DoD
  and DoE nuclear activities under DoD
  – Concepts for resolving major issues				

• If analysis warrants, realign U.S. nuclear infrastructure under Secretary of Defense				

Figure S.1—Recommended Next Steps

The ultimate goal should be to provide focus and leadership of the enduring nuclear-weapons-related functions. This process should begin by laying out the known problem areas and potential cost savings associated with this consolidation and characterizing the intellectual challenges of the next several decades that are better addressed within a new and streamlined national nuclear infrastructure.

Finally, we recommend a consolidation of another kind. Our review noted the number of separate DoD, DoE, Office of Technology Assessment (OTA), General Accounting Office (GAO), and other studies dealing with aspects of the nuclear infrastructure problem. A useful early step in moving toward consolidation options would be to organize an activity bringing the study groups together to facilitate an effective interchange of viewpoints and ideas.

GLOSSARY

AAT	Accident Advisory Team
AC&TM	Arms Control and Threat Management
AEC	Atomic Energy Commission
AFRRI	Armed Forces Radiological Research Institute
AFSWP	Armed Forces Special Weapons Project
AGT	Aboveground Testing
ALCM	Air-Launched Cruise Missile
ARPA	Advanced Research Projects Agency
ATSD (AE)	Assistant to the Secretary of Defense (Atomic Energy)
BDA	Bomb Damage Assessment
BMDO	Ballistic Missile Defense Organization
BUR	Bottom-Up Review
CINC	Commander-in-Chief
CINCSTRATCOM	Commander-in-Chief Strategic Command
CONUS	Continental United States
CTBT	Comprehensive Test Ban Treaty
CTR	Cooperative Threat Reduction
DASA	Defense Atomic Support Agency

DDR&E	Director, Defense Research and Engineering
DNA	Defense Nuclear Agency
DoD	Department of Defense
DoE	Department of Energy
DoE/DP	Assistant Secretary of Energy for Defense Programs
DSB	Defense Science Board
DSWA	Defense Special Weapons Agency
EMP	Electromagnetic Pulse
ERDA	Energy Research and Development Administration
FSU	Former Soviet Union
GAO	General Accounting Office
HAC	House Appropriations Committee
HE	High Explosive
IACRO	Interagency Cost Reimbursement Order
IAEA	International Atomic Energy Agency
ICBM	Intercontinental Ballistic Missile
JCS	Joint Chiefs of Staff
JSTPS	Joint Strategic Targeting Planning Staff
LANL	Los Alamos National Laboratory
LLC	Limited Life Component
LLNL	Lawrence Livermore National Laboratory
MILCON	Military Construction
MIPR	Military Interdepartmental Payment Request
MRC	Major Regional Contingency
NEST	Nuclear Emergency Search Team

NOE	Nuclear Weapons Operational Employment
NPR	Nuclear Posture Review
NPT	Non-Proliferation Treaty
NRL	Naval Research Laboratory
NTS	Nevada Test Site
NWC	Nuclear Weapons Council
NWE	Nuclear Weapons Effects
NWTI	Nuclear Weapons Technical Inspection
O&M	Operations and Maintenance
OSD	Office of the Secretary of Defense
OSIA	On-Site Inspection Agency
OSTP	Office of Science and Technology Policy
OTA	Office of Technology Assessment
PDA	Procurement Defense Agency
PDP	Program Decision Package
PRP	Personnel Reliability Program
QA	Quality Assurance
R&D	Research and Development
RDT&E	Research, Development, Test, and Evaluation
RF	Radio Frequency
RT	Reliability Testing
SAC	Senate Appropriations Committee
SAGE	Scientific Advisory Group on Effects
SASC	Senate Armed Services Committee
SDIO	Strategic Defense Initiative Office
SEAB	Secretary of Energy Advisory Board
SETA	Systems Engineering Technical Assistance

SNL	Sandia National Laboratory
SNM	Special Nuclear Material
SPO	System Program Office
START I	Strategic Arms Reduction Treaty (July 1991)
START II	Strategic Arms Reduction Treaty (January 1993)
STRATCOM	Strategic Command
UGT	Underground Testing
UNSC	United Nations Security Council
USG	United States Government
WMD	Weapons of Mass Destruction

Chapter One

INTRODUCTION

BACKGROUND

The end of the Cold War had an enormous effect on the national security structure of the United States. All of the military services, the Joint Staff, and the Department of Defense began a thorough analysis of national security strategy and its supporting apparatus. The role of nuclear weapons has been an important part of this analysis. For 40 years, nuclear weapons—in ever-increasing numbers—played a central role in the U.S. national security strategy. With the collapse of the Soviet Union, the signing of the START II treaty, and the subsequent reduction (and, in the case of the Soviet Union, the dispersal of some of its weapons to former member states), the future role of U.S. nuclear weapons and their entire supporting infrastructure has come under close scrutiny.

The nuclear weapon infrastructure of the United States is large, complex, and expensive. It resides not only in the military services and the Department of Defense (DoD), but also across the Department of Energy (DoE) and a host of commercial organizations. The cost of such a large infrastructure was supported by Congress while the United States faced a large nuclear threat from the former Soviet Union. But now that the threat has declined, the time is ripe for a reassessment and reorganization of that infrastructure, with a major reduction in costs.

As one part of that reassessment, this report responds to a congressional request for an independent examination of options to accomplish the functions now performed by the Defense Nuclear

Agency (DNA). This request was motivated by a desire to resolve two quite different perspectives about the future of DNA. One perspective would eliminate DNA and assign its current functions to other organizations: the military services, the Advanced Research Projects Agency (ARPA), or the DoE. The other perspective would continue DNA as an entity, but with an expanded role as a DoD focal point for activities relating to all types of "special weapons"—chemical and biological weapons of mass destruction (WMD), as well as nuclear weapons and special-purpose conventional weapons. Both perspectives reflected a desire to reduce the cost and size of at least one portion of the nuclear weapon infrastructure and refocus its activities on the problems of the post-Cold War era, but there were major disagreements in Congress about how to proceed. Against this background, the Congress requested a study examining a range of options, some of which it specified. The result was a four-month RAND study and the current report, which evaluates the congressional options and suggests others for consideration.

STUDY OBJECTIVE AND SCOPE

The original objective of this study was to respond to the congressional tasking contained in Public Law 103-139, November 1993, to assess five specific options for the future conduct of DNA's current functions:

- Transferring DNA's functions to the individual armed services or ARPA;

- Maintaining DNA as a separate agency, evolving to meet conditions of the new national security environment;

- Transferring DNA's functions to the DoE nuclear weapon laboratories;

- Combinations of any of these options;

- Reorganizing DNA to reduce its operating, management, administrative, and other overhead costs significantly.

DNA, however, is only a small part of the entire U.S. nuclear infrastructure, and it has important functional interactions with many of the other parts. In the course of addressing these options, it therefore quickly became apparent that they reflected only one fairly small aspect of a larger, more complex problem: the future direction, organization, and management of the entire U.S. nuclear infrastructure. It became clear that an evaluation of alternatives for accomplishing DNA functions should be set not only in the emerging international security environment requested by Congress, but also in the context of that larger problem. This led to a broadening of the scope of the study, to include a second objective: to address the national nuclear weapon infrastructure in its entirety—identifying the larger issues regarding its future evolution and suggesting organizational approaches to resolve those issues (including the management of DNA's current functions). This in turn led to the consideration not only of the congressionally specified options, focused on DNA, but also of some broader options concerned with the overall infrastructure.

These two objectives define the scope of the study insofar as *breadth* is concerned. The *depth* of the study was defined by its four-month duration: i.e., in some areas we were able only to identify and illuminate issues and suggest solutions, not perform completely definitive analysis.

STUDY APPROACH

We began the study by identifying a range of plausible future international security environments, the role that nuclear weapons would (or could) play in these environments and in U.S. national security strategy, and the resulting enduring requirements imposed on the U.S. nuclear infrastructure. This analysis (presented in Chapter Two) drew on recent studies of the various roles that nuclear weapons could play in the post-Cold War era. We used this set of key enduring nuclear infrastructure requirements as an organizing device for our subsequent analysis.

We then turned to DNA and developed a profile of its current functions and the manner in which they are carried out. In the process of developing this profile, we conducted extensive interviews within organizations and with individuals across the spectrum of U.S. nuclear weapon activities.[1]   The resulting profile identifies four primary functional areas into which DNA's current activities fall. For each of these areas, we developed an understanding of the role that DNA plays, what it does and how it goes about it, the approximate scale of its efforts, and issues regarding future directions. We also developed a general perspective on the changes currently under way within DNA. This profile (presented in Chapter Three) served as the point of departure for our analysis of the congressional options.

We then turned to the congressional options, assessing: (1) the risks, if any, each of them posed to the maintenance of essential functions associated with enduring nuclear infrastructure requirements; and (2) the likely cost effect—to the extent we could ascertain it— associated with the adoption of a particular option. The results of these assessments are presented in Chapter Four.

We then turned our attention to the broader subject: the problems of downsizing and refocusing the entire U.S. nuclear infrastructure in the new post-Cold War environment. We developed a perspective on the evolution of the nuclear infrastructure over the last 50 years, its current state, and the changes it is undergoing as it adapts to post-Cold War priorities.[2]   From this perspective, we identified a set of principles for shaping the future evolution of the infrastructure. These principles were guided by the enduring requirements we identified previously and by the stresses we saw challenging the maintenance of those requirements in the post-Cold War world. We used these principles to develop two options beyond those on the congressional list. These involved, first, the consolidation of all nuclear infrastructure activities currently within DoD and, second, the consolidation within DoD of the entire U.S. nuclear infrastructure. Although we had insufficient data and time for a full analysis, we identified a number of advantages associated with these options. We also identified a number of key questions that would determine their ultimate feasibility and value. All of this is presented in Chapter Five.

---

[1]Table 1.1 identifies the key organizations contacted. A comprehensive list of contacts made (over 30 organizations and numerous knowledgeable individuals) and briefings received is contained in the bibliography.

[2]In developing this perspective, we benefited greatly from the extensive interviews we conducted within organizations and with individuals across the entire spectrum of U.S. nuclear weapon activities.

Table 1.1

Key Organizations Visited in the Study

Department of Defense	**Congress**
ATSD(AE)	Six congressional committees
USD(A)	
USD(P)	**Department of Energy**
Deputy Secretary of Defense	Assistant Secretary for
	Defense Programs
Services	Weapons laboratories
USA/USN/USAF	Los Alamos
	Lawrence Livermore
Defense Agencies	Sandia
Defense Nuclear Agency	
Advanced Research Projects Agency	**Other Government Organizations**
Defense Intelligence Agency	National Security Council
On-Site Inspection Agency	NIC
	Central Intelligence Agency
Other Defense Organizations	Department of State
Office, Joint Staff	
Unified and Specified Commands	**Primary Contractors**
	Selected from DNA's top ten
	R&D contractors

Finally, we formulated a number of findings and recommendations, some focused on the congressional options, some addressing the broader issues associated with the entire U.S. nuclear weapons infrastructure, based on the totality of this analysis. These are presented in Chapter Six.

SIGNIFICANCE OF STUDY

The report's ultimate value is in redefining the issue and proposing a serious examination of the most comprehensive consolidation options. They, by contrast with the originally proposed options, could achieve massive cost savings while assuring coherence in the residual nuclear weapons effort.

STRUCTURE OF THIS REPORT

As indicated, Chapter Two discusses the future national security environment facing the United States, Chapter Three provides a profile of DNA, and Chapter Four assesses the congressional options for the future of DNA. Chapter Five discusses the broader issues involving the entire U.S. nuclear weapons infrastructure, and Chapter Six presents the findings and recommendations of the study. The appendix provides information on recent actions Congress has taken regarding DNA, providing a background to the current study. The bibliography lists all the source material we consulted during the course of the study: publications, briefings, letters and memorandums, and meetings within organizations and with individuals.

Chapter Two

THE FUTURE NATIONAL SECURITY ENVIRONMENT

The end of the Cold War has had its most profound effect in the area of nuclear weapons. Dramatic changes are under way in the national security role of nuclear weapons, the size of the U.S. nuclear arsenal, the nation's future needs for nuclear weapons and the infrastructure supporting those weapons, and (perhaps most important) the source and nature of future nuclear threats to the United States. Given this setting, it became clear early in the problem formulation phase of the study that an evaluation of alternatives for accomplishing DNA functions should be set within a broad context dealing with two important aspects of the future environment. These are (1) the emerging national security environment, which sets the required level of capabilities for our weapons; and (2) the evolution of the U.S. nuclear (and related) infrastructure, which provides a backdrop for setting priorities for management and resource allocation. We discuss the first of these in this chapter. The second is the subject of Chapter Five.

THE RANGE OF POSSIBLE NUCLEAR FUTURES

Predicting the future course of international events in the post-Cold War era is a process fraught with uncertainties. Table 2.1 lists key uncertainties in some of the most important factors shaping the future world, insofar as nuclear weapons and other weapons of mass destruction (WMD) are concerned. How uncertainties such as these play out will largely determine:

- The role that nuclear weapons play in this future world, and in U.S. national security strategy.

Table 2.1

Some Key Uncertainties About the Future International Security Environment

Factor	Uncertainties
Russia	If Russia transitions to a highly nationalistic and more authoritarian state, will it also reopen a strategic nuclear competition with the United States, especially if it can no longer afford vast and modern conventional forces? Will the United States find its nuclear force structure useful in helping to extend deterrence to some of Russia's neighbors?
China	Will China seek a nuclear arsenal much closer in size to that of Russia and the United States?
Middle East	Will coalitions emerge and present unified fronts to the West in key regions (the Persian Gulf, North Africa) or contexts (WMD arms control)?
Regional hegemons	Will aggressive would-be regional hegemons (e.g., North Korea) emerge brandishing WMD and challenging the status quo in regions of vital U.S. interest?
Nuclear proliferation	Will the number of nations with at-the-ready nuclear arsenals or preplanned virtual nuclear arsenals continue to grow or recede? Will either Japan or Germany feel compelled at some future time to acquire nuclear weapons? Will Russia and China help or hinder the spread of modern weapons and delivery systems?
High-technology diffusion	Will the military technology that currently provides the United States with a technical advantage on the battlefield (e.g., Desert Storm) be diffused to other nations? If so, how soon and how fast?

- The source and nature of future nuclear threats to the United States.

- The future U.S. requirements for nuclear weapons and supporting infrastructures.

Using just these uncertainties as a set of free variables, one can derive a vast number of future worlds, some more or less likely, but all of them plausible. Important metrics defining and describing these future worlds, insofar as nuclear weapons are concerned, include (Molander and Wilson, 1993):

1. The number and names of nations that maintain nuclear arsenals at the ready.

2. The size of these at-the-ready nuclear arsenals.

3. The restrictions, if any, that might be imposed on virtual nuclear arsenals.[1]

4. The declaratory policies existing in each of the various nuclear states on the role of nuclear weapons.

5. The size and character of defenses (if any) against nuclear attack.

A number of authors have developed such future worlds.[2] For example, in a study of the possible courses of nuclear proliferation in the post-Cold War era, Molander and Wilson (1993) present four "illustrative alternative asymptotes," hypothetical future worlds towards which the "real" world could evolve over time. These four representative future environments are as follows:[3]

Menacing: a highly proliferated world with a great deal of nuclear "disorder" and few "rules of the nuclear road," save possibly a perpetuating (if successful) cultural taboo on nuclear use that relies, *inter alia,* on an expanding web of bipolar or multipolar international deterrence relationships to keep the nuclear peace.

Increasing risk: acceptance of an inexorable slow growth in the number of nuclear-armed states with new members of the "club" grudgingly (or sometimes willingly) integrated into the existing nuclear order and carefully educated to a set of nuclear "norms" of behavior and associated deterrence/balances concepts.

Stable risk: a two-tiered international system, in which a handful of "haves" maintain substantial (but limited by treaty) "at-the-ready" nuclear arsenals and commit themselves individually or collectively

---

[1] By "virtual nuclear arsenals" we mean arsenals that by virtue of prior planning can be built or assembled in a (strategically) short period of time.

[2] See, for example, McNamara (1993), Molander and Wilson (1993), Millot et al. (1993a, 1993b, 1993c), Warnke (1994), Buchan (1994), and Martel and Pendley (1994).

[3] For greater clarity, we have renamed Molander and Wilson's four worlds. Their original terminology was: I. "High entropy" deterrence; II. An ever-slowly-expanding nuclear club; III. A two-tiered static "have-a-lot/have none" international system; and IV. Nuclear de-emphasis.

(most likely through the U.N. Security Council (UNSC)) as explicitly as necessary to maintaining the security of the "have-nots."

Benign: a world in which nuclear weapons have been greatly deemphasized and existing "at-the-ready" nuclear arsenals virtually eliminated (a handful of states maintain tens to hundreds of nuclear weapons "at-the-ready") underwritten by an unprecedented, comprehensive, and highly intrusive international inspection and collective enforcement regime.

Table 2.2, an extension of the work of Molander and Wilson, provides further details of the characteristics of these four alternatives.[4] The dominant distinguishing characteristic of these four nuclear environments (and the rationale for the sequence in which they are arrayed) is the degree of nuclear "order"—from a low degree of nuclear order (i.e., a high degree of disorder) in the first case, Menacing, to a very high degree of order in the last case, Benign.

The reader can hazard a guess as to which of these future worlds is the most likely. Over the next few decades, the real world may in fact tend first toward one of these worlds, then another. Taken together, these four asymptotic worlds of Molander and Wilson appear to bound the range of plausible nuclear possibilities. This is not to say that some environment more extreme than these could not occur. But these four seem to cover the most likely range.

---

[4]The numbers provided in Table 2.2 are intended to be strictly notional and illustrative. In the first three cases—Menacing, Increasing Risk, and Stable Risk—the numbers presented for at-the-ready arsenals are a logical extrapolation from current circumstances, under the assumptions that characterize these end states. For example, in the Menacing case, Ukrainian retention of a significant nuclear arsenal would presumably lead to both the United States and Russia retaining arsenals closer to the START I numbers than to the START II numbers, and to some nuclear buildup by China. India and Pakistan are already credited with weapon numbers in the tens with no near-term prospect of a negotiated ceiling on their arsenals: hence the estimate in the first two cases that they would continue to build to larger numbers akin to those currently estimated for Israel. In the final Benign case, the numbers shown for at-the-ready arsenals are rough estimates of what the international community might agree is appropriate to deter breakout from such a far-reaching agreement. (In developing these numbers, we were guided by the data in Norris et al. (1994) and Albright et al.(1993).)

Table 2.2

Some Illustrative Alternative Future Nuclear Environments

Plausible Nuclear Asymptotes (End States)	At-the-Ready Arsenals[a]		Other Key Characteristics
Menacing	U.S./Russia	5000+	Weak international
Highly "nuclear proliferated	China/Ukraine	1000+	security system
world"	U.K./France	100s	Mixed national
Many at-the-ready and virtual	Israel/India/		counterproliferation
nuclear arsenals	Pakistan	100+	responses (defenses,
Few rules of the nuclear	10+ nations	10s	counterforce, power
road			projection adaptation, or
Bilateral and multilateral			neo-isolationism.
deterrence relationships			
Increasing Risk	U.S./Russia	3000+	No major change in current
Inexorable growth in number	China/Ukraine	1000+	international security
of nuclear-armed states	U.K./France	500+	system
New nuclear "club" members	Israel	200	National counterproliferation
integrated and educated	India/Pakistan	100+	responses constrained by
on norms of behavior	10+ nations	10s	arms control
Little improvement in Non-		to 100s	
Proliferation Treaty (NPT)			
inspection regime			
Stable Risk	U.S./Russia	2000	Strengthened international
Two-tiered international	China	600	security system with new
system	U.K./France	400	security guarantees
Idealized NPT "Haves"	Others	0	Improved NPT regime
(presumably UNSC			(inspection and
PermFive) maintain			enforcement)
substantial but treaty-			Counterproliferation deem-
limited arsenals			phasized
Significantly improved			
International Atomic			
Energy Agency (IAEA) in-			
spection regimes			
Benign	U.S./Russia	300	Robust international security
"Virtual abolition" of	China	200	system (great powers agree
nuclear arsenals and	U.K./France	100	on "nuclear order") compre-
missions	Others	0	hensive/ intrusive/ relentless
A few states maintain up to			inspection and enforcement
100s of at-the-ready			regime
nuclear weapons to deter			Robust counterproliferation
breakout			investment

[a]Estimated numbers of nuclear weapons.

There is nothing magic about these four worlds or about the numbers in Table 2.2. The important point about these possible nuclear futures—or similar sets derived by other authors—is not any specific number or set of numbers, but *the wide range of possible numbers,* of nuclear weapons and of countries possessing them, from a "benign" extreme of a few hundred nuclear weapons worldwide with very little proliferation, to the "menacing" extreme of many thousands of weapons worldwide and widespread proliferation. In view of the many uncertainties facing us regarding the future, it is prudent to assume that the benign extreme is the least likely of the four.

The United States would be making a mistake to place any important bets today on a future world having only a few nuclear weapon states with only tens to hundreds of nuclear weapons each.

IMPLICATIONS FOR NUCLEAR WEAPONS AND FOR THIS STUDY

One thing seems clear from this analysis: Nuclear weapons will not go away. They may have become much less important in political, diplomatic, and military calculations than they were at the height of the Cold War, the size of nuclear arsenals throughout the world may undergo continued sharp declines, and the threat posed to the U.S. homeland may be sharply reduced, but few, if any, of the nations that currently (1994) possess nuclear weapons are likely to give them up, and additional actors on the world stage may well acquire such weapons. Some of these new nuclear actors may be less stable, less responsible, and less easily or reliably deterred than the nuclear powers were during the Cold War, and one or more nuclear weapons just might go off in anger over the next few decades. Indeed, it is not difficult to imagine situations in the post-Cold War era in which the actual use of nuclear weapons *is more probable* than it was during the Cold War. Further, although it may seem difficult to imagine today, it is plausible that the United States may need to rediscover a version of extended deterrence if Russia or China should again become aggressive or if a new nuclear actor appears on the world stage.

Although nuclear weapons will be less important, they will therefore continue to have a role in international affairs, and that role will be vastly different than it was during the Cold War. In this uncertain post-Cold War era, the United States must have nuclear forces that are qualitatively superior to those of any challenger.

There is one "future" that the United States must avoid at all costs: one in which the United States has placed itself at a

nuclear disadvantage relative to any other country. Indeed, if the price is low enough, there may well be value in maintaining current U.S. nuclear advantages, however uncertain the future or applicability of such advantages might currently seem.

ENDURING REQUIREMENTS

The United States will have a continuing need for some nuclear weapons, albeit many fewer than during the Cold War, and a nuclear weapon infrastructure, albeit much smaller than during the Cold War. There will be continuing requirements:

1. To maintain the U.S. nuclear stockpile. Stockpile stewardship—the safety and security of nuclear weapons, and the competencies required to carry out these functions—will be required for the foreseeable future, in all foreseeable future worlds.

2. To maintain at least a minimal capability to modernize that stockpile, to replace aging weapons or meet changing requirements (e.g., safer weapons and weapon materials). Maintenance of the core physics and engineering competencies involved in nuclear weapon design and fabrication will be required for the foreseeable future, once again in all foreseeable future worlds.

3. To maintain the ability to employ nuclear weapons effectively, and to conduct military operations in a nuclear environment, if either of these should become necessary at some future time. This involves a number of core competencies in nuclear weapon effects, including:

    • An understanding of the environments produced by nuclear weapons.

    • An understanding of the interactions of these environments with military systems.

    • An understanding of the response of those systems to these interactions.

    • The ability, through simulation and testing, to determine the nuclear hardness of friendly systems and to assess the vulnerability of foreign systems.

The necessary knowledge bases and capabilities in these core competencies must be preserved, for a future time when the United States might need to use them. This is true in all of the foreseeable future worlds, but even more so in the more menacing ones.

4. To maintain a capability to reconstitute a larger nuclear weapons inventory, if future international developments were to make that necessary. How *many* additional weapons, and how quickly, are key open questions here. This is primarily a surge production issue and a critical nuclear materials inventory issue, not a core competencies issue.

5. To maintain all of the above in an extended era of no underground nuclear testing. This represents a major challenge, which the United States has never had to face before.

6. To maintain a capability to pursue effective nuclear (and other WMD) threat reduction and arms-control activities. Regardless of which nuclear future obtains, the United States should continue efforts to reduce the threat posed by nuclear and other weapons of mass destruction. Part of that reduction could involve arms-control verification; the technologies associated with such verification are highly specialized and need to be retained.

These enduring requirements—the functions the United States must be able to perform—are the context in which the future course of DNA must be assessed.

Chapter Three

A PROFILE OF THE DEFENSE NUCLEAR AGENCY

INTRODUCTION

Having described the future international security environment, this chapter focuses on DNA itself. DNA is a complex organization with diverse responsibilities. Its organization and missions have evolved over the past 47 years as the external threat environment and defense priorities have changed. At present, DNA is an organization in transition, from its Cold War, primarily nuclear role to some new, mixed role in the post-Cold War era.

As currently constituted, DNA activities fall into four primary areas: nuclear weapons stockpile support, nuclear weapons effects research and operational support, nuclear threat reduction and arms control, and nuclear-weapons-effects-related conventional defense technologies. The first three of these areas contribute directly to three of the enduring nuclear infrastructure requirements identified in Chapter Two:

- Nuclear weapons stockpile support: maintenance of the U.S. nuclear stockpile.

- Nuclear weapons effects research and operational support: maintenance of the ability to employ nuclear weapons effectively and to conduct nuclear operations in a nuclear environment if either of these should become necessary at some future time.

- Nuclear threat reduction and arms control: maintenance of the ability to pursue effective nuclear (and other WMD) threat reduction and arms-control activities.

The fourth area, conventional defense technologies, represents an extension of DNA's activities into a variety of nonnuclear military technology areas—areas in which DNA can exploit capabilities originally developed as part of its nuclear responsibilities.

This chapter discusses each of these four areas in turn, describing the role that DNA plays in each area, what DNA does and how it goes about it, and the approximate scale of the efforts in terms of budget. We also make a number of observations in each area and identify issues regarding future directions. The later part of the chapter makes a number of observations regarding DNA as a whole.

FUNCTIONAL AREAS

Nuclear Weapons Stockpile Support

1. What is this functional area, and what are its future needs? This functional area directly supports the maintenance of the U.S. nuclear stockpile. It includes those procedures, systems, and supporting technologies required to ensure the safety, security, reliability, and survivability of the stockpile. As discussed in Chapter Two, there will be continuing need for these functions for the foreseeable future in all likely future worlds.

2. What does DNA do? DNA is one of several organizations contributing to the support of the nuclear weapons stockpile. DNA's primary role is in the management and auditing of stockpile support activities associated with that portion of the warhead stockpile under DoD custody. DNA's activities include the development and implementation of specific procedures, systems, and supporting technologies to ensure the safety, security, reliability, and, to a limited extent, survivability of the nuclear weapons stockpile within DoD's custody (U.S. DoD directive, "Defense Nuclear Agency," 1993).

DNA addresses operational safety and security issues in training programs (e.g., the Nuclear Weapons School), in studies (e.g., Minuteman III quantitative analyses), and in independent field inspections of the services' nuclear weapons operational procedures and storage facilities. In addition to developing and operating the computer database and reporting systems for the Joint Chiefs of Staff

(JCS) that track and document the status of all nuclear weapons under DoD's control, including their location, movement, and service actions, DNA maintains a site folder on all U.S. nuclear storage facilities worldwide and can transmit electronically appropriate data and images to field teams that might be called on to react to an accident or security threat. DNA also maintains a command center and response teams to react to a nuclear accident or incident anywhere in the world.

DNA supports DoE's technical assessments of nuclear weapon physical safety through coordinated efforts, in which DNA develops and analyzes normal and stressful operational environments (e.g., accidents) and DoE assesses their effect on the safety and proper functioning of the warheads.

Reliability of the nuclear stockpile is primarily a DoE responsibility, but DNA supports DoE's reliability testing and assessments by keeping detailed maintenance records and failure reports. DNA also supports DoE's reliability assessments by specifying and monitoring operational environments associated with the stockpile-to-target sequence (e.g., missile warhead storage/basing environments and in-flight conditions). DNA also oversees the DoE quality assurance program and manages the limited life cycle spares program, which supplies parts made in the DoE to the DoD.

Survivability of the nuclear weapons stockpile (as well as the delivery platforms and support communications and intelligence systems) is also a DNA concern. Although the technical assessments are a DoE laboratory function, DNA supports the laboratories by developing and operating shared test facilities. In the past, these included underground nuclear testing (UGT) facilities at the Nevada Test Site (NTS), but, with the underground testing moratorium and the likelihood of a comprehensive test ban, DNA's focus has turned to nonnuclear aboveground testing (AGT) using simulators developed as part of its nuclear weapons effects research activities.

3. How does DNA do it, and what are the costs? DNA largely conducts these efforts using in-house (primarily service) personnel assigned to the DNA Field Command. In FY93 (the latest year for which complete financial data are available), DNA spent about $10 million[1] on these activities.[2] This level has been relatively stable over recent years and is unlikely to change dramatically in the immediate future.[3]

4. Observations and issues. Stockpile support is obviously an important and complex mission for which the United States will have a continuing need. Because of the potential grave consequences arising from an accident or misuse, prudence dictates that the nuclear community exercise extreme diligence in its custody of these weapons and in its efforts to ensure the continuing viability and credibility of our nuclear deterrent. DoE builds and maintains weapons with an emphasis on safety and security, but operational safety and security is up to the service custodians. The services and nuclear CINCs manage and inspect their nuclear operations carefully, but DNA provides an independent oversight and audit that, considering the stakes, is essential.

Nuclear Weapons Effects Research and Operational Support

1. What is this functional area, and what are its future needs? This functional area provides the detailed understanding of NWE required to support the effective employment of nuclear weapons by U.S. military forces, and the conduct of military operations by those

---

[1]The DNA RDT&E resource allocation and financial management system is organized in a hierarchical structure with "Projects" (the highest level, of which there are eight in the RDT&E area), "Program Decision Packages" or "PDPs" (the next level down, of which there are about 220), and "Work Units" (the third level down, of which there are about 2100). This and subsequent statements in this report concerning financial aspects of DNA activities are based on a RAND analysis of the FY93 DNA budget and expenditures at the PDP level.

[2]This includes activities in two DNA-designated areas—stockpile management and accident/incident response (DNA, Field Command, FY 94)—but does not include the cost of any simulator testing to validate stockpile survivability. We have included this in the next functional area—nuclear weapons effects and operational support.

[3]The U.S. nuclear weapon stockpile is in the midst of a major drawdown. Ultimately, once the stockpile has stabilized at a level much smaller than that at its Cold War peak, it should be possible to significantly reduce the overall U.S. expenditures (by DNA, the rest of DoD, and DoE) for stockpile support activities. During the current transition period, however, the many activities associated with the removal of warheads from the operational inventory, their transfer via the DoD depot system to DoE facilities, their subsequent dismantling, etc., and *the safety and security of these warheads at each step in this process* make such funding reductions difficult to achieve.

forces in a nuclear (or potentially nuclear) environment. It includes the use of NWE information: (1) to support nuclear lethality assessments (for targeting and collateral damage assessments) and the development of detailed targeting plans for the employment of nuclear weapons, (2) to assess the nuclear vulnerabilities of U.S. military systems, (3) to develop hardening technologies and design approaches that mitigate the effects of nuclear weapons and enhance the survivability of U.S. systems, (4) to develop and operate test facilities to validate the nuclear hardness of friendly systems (new and modified) and assess the vulnerabilities of foreign systems, and (5) to support CINCs in planning to fight nuclear adversaries in a major regional contingency (MRC).

As discussed in Chapter Two, there is little likelihood that U.S. military forces will have to employ nuclear weapons or conduct military operations in a nuclear environment in the immediate future. However, the possibility exists, and over the long term, geopolitical developments could lead to situations in which nuclear use by a proliferant is deemed more likely than was a NATO-Warsaw Pact conflict during the Cold War. This means that there is a clear need to preserve the NWE knowledge base that the United States now has for a future time when the United States might need to use it. In addition, future nuclear "situations," if and when they arise, are likely to involve scenarios quite different from those of the Cold War (e.g., an MRC involving U.S. forces confronting those of a "rogue state" possessing several nuclear weapons). Because of this, future U.S. systems may have to operate and survive in distinctly different types of (possibly nuclear) environments. We also should preserve the capabilities to do some testing, as well as merely "knowledge" required to understand system survivability in these new environments, if and when the need arises.

2. What does DNA do? DNA has been the DoD center of excellence for understanding and documenting the effects of nuclear weapons on systems for both vulnerability/hardening and targeting support to the services and CINCs. It is the centralized DoD focus for this nuclear research and is an advocate for nuclear issues. DNA has been the leader in developing hardening technologies and design approaches that mitigate the effects of nuclear weapons and in developing and operating test facilities to validate the nuclear hardness of friendly systems and to assess the vulnerabilities of foreign systems

(U.S. DoD Directive, "Defense Nuclear Agency," 1993). DNA's nuclear effects research and testing can be broken down into three areas: (1) nuclear weapon environments, (2) interactions of these environments with the systems, and (3) the response of the systems to these interactions.

In the research area, DNA characterizes the environments produced by nuclear weapons (e.g., blast, thermal, radiation, and electromagnetic pulse (EMP)) in various operational regimes (e.g., underwater, buried, ground burst, air burst, high-altitude burst, and bursts above the sensible atmosphere). It develops analytical models and computer codes based on physical principles and empirical data from nuclear tests and simulations. These models cover all effects and include, as just one of many examples, the propagation of radio waves through ionospheric disturbances caused by high-altitude nuclear bursts. Similarly, DNA models and tests the response of systems to these environments by using codes and effects tests.

DNA develops nonnuclear systems to simulate specific nuclear threat environments, to better understand nuclear environments, and to test systems exposed to those environments. These simulators range from small-scale conventional explosives tests to large-scale blast and thermal generators to EMP and X-ray generators. The nuclear UGT ban has presented a significant challenge in this regard: the design of nonnuclear AGTs to cover adequately the nuclear environment spectrum for system-level (versus subsystem or component) testing.

DNA is also deactivating and cleaning up nuclear test sites and is implementing plans to support renewed UGTs if required.[4] DNA also manages the National Nuclear Test Personnel Review and the DoD Human Radiation Research Review (U.S. DoD Directive, "Defense Nuclear Agency," 1993).

In operations support, DNA assists the U.S. Strategic Command/ Joint Strategic Target Planning Staff (USSTRATCOM/JSTPS) (and other nuclear CINCs) in nuclear employment planning by: (1) assessing the vulnerability of enemy systems to support targeting

---

[4]Safeguard C, the requirement to maintain an atmospheric test readiness posture, was canceled recently.

decisions, (2) assessing the survivability of U.S. systems to support posture decisions and operational employment planning (and to identify deficiencies to be addressed in new systems or systems modifications), and (3) developing adaptive planning tools (e.g., software models) for CINC staff use.

The DNA investment in environment research has been diminishing, since such research is fairly mature and well documented; the same is true for the system response modeling and testing area, since this involves system-specific efforts and there are currently few new systems with nuclear hardness specifications being acquired. The recent shift in emphasis at DNA has been away from traditional strategic-force issues (e.g., missile silo hardness) toward emerging issues such as the survivability and operability of space-based C4I and missile defense systems.

3. How does DNA do it, and what are the costs? DNA out-sources almost all of its efforts in this area, relying on the contractor community and other government laboratories for the conduct of most of these activities. DNA maintains a core of government technical managers to set priorities, establish research objectives, monitor the quality of the work, and provide continuity and corporate memory. In FY93, DNA allocated about $413 million to this area.[5] The expenditure numbers for FY94 and later years are expected to be considerably lower.[6]

4. Observations and issues. With the end of the Cold War, the suspension of underground nuclear testing, the reduction of the number of nuclear weapons in the stockpile, and the changes in U.S. military strategy and operational planning, the level of activity in this area is being sharply reduced. At the same time, a debate has begun regarding the future need for NWE research. Does the United States already know most (or all) of what it needs to know about nuclear weapons effects to meet the needs of the post-Cold War era? Or are

---

[5]This includes activities in eight DNA-designated areas: facility survivability, systems lethality, advanced radiation simulation, core competency for future initiatives, adaptive targeting, operability in nuclear disturbed environments, hard target interactions, and computing and information systems (DNA, 1994 Program).

[6]The FY94 DoD Appropriations Bill reduces Program Element 62715H, which covers DNA's RDT&E activities in this area, by about 40 percent from the FY93 figure (see the appendix).

there important areas requiring continued investigation (e.g., understanding how new systems will operate in a nuclear weapons environment)?

There is no universal agreement on the answers to these questions. However, judging by our prior knowledge and interviews with other knowledgeable experts, it appears that the United States does know most of what it needs to know about nuclear weapons effects and that the primary task *for the immediate future* is the preservation and application of this knowledge (for a future time when the United States might need to use it), not the creation of new knowledge. This said, technology changes rapidly and the effect of nuclear environments on new devices or systems is not the same as it was on older technologies, even though the radiation environments may be the same. This is perhaps even more important than before as the United States comes to depend more on commercial systems for information and intelligence dominance. Without continued testing and expertise, the United States could encounter important surprises (and be unable to cope with them well) if nuclear weapons were used in a future contingency.

As indicated above, a future time could well arrive—e.g., if the United States is confronted by a rogue state armed with nuclear weapons—when the creation of new physics knowledge is required. For example, this could involve the vulnerabilities of specific U.S. systems in nuclear environments, different from those expected during the Cold War, that might occur during a conflict involving that rogue state. This suggests that the United States must not only preserve current NWE knowledge but must also retain *the capability to develop new knowledge in the future* if that becomes necessary.

Taken together, our observations imply the preservation of some minimal level of NWE simulation and testing capability, along with the current knowledge base and the testing of any commercial hardware that may be used in military systems.

Nuclear Threat Reduction and Arms Control

1. What is this functional area, and what are its future needs? This functional area involves: (1) a variety of activities supporting ongoing U.S. nuclear threat reduction efforts, including most prominently those associated with the dismantling of FSU nuclear weapons within the context of the Nunn-Lugar Cooperative Threat Reduction (CTR) legislation and (2) R&D activities supporting compliance monitoring of arms-control treaties, currently including those regulating underground nuclear testing, strategic and intermediate-range nuclear force reductions, chemical weapons elimination, and conventional force reductions.

As indicated in Chapter Two, these activities are of increased importance in the post-Cold War era. Regardless of which nuclear future prevails, the United States will almost certainly continue to pursue efforts to reduce the threat posed by nuclear and other weapons of mass destruction. In the near term, taking advantage of the "window of opportunity" now available to encourage and facilitate the dismantling of FSU nuclear weapons takes priority, along with CTR activities supporting that effort. Over the longer term, arms-control verification will be a continuing need; many of the technologies associated with such verification are highly specialized and must be retained.

2. What does DNA do? DNA's role in this area has recently undergone considerable growth, which is expected to continue: in nuclear threat reduction because of Nunn-Lugar CTR activities, and in arms-control verification because of the increasing number and scope of treaties in recent years.

In CTR, which is a legislated responsibility of OSD, DNA acts as the OSD executive agent for the implementation of support agreements that have been negotiated between the United States and FSU states on the safe and secure dismantling of FSU nuclear weapons. In this capacity, DNA is involved in a variety of coordination and procurement functions, with procurement of relatively routine equipment used in dismantling FSU nuclear weapons and transporting, storing, or securing sensitive components accounting for the bulk of the expenditures. As of February 1994, DNA had executed 32 CTR projects: 13 with Russia, 8 with the Ukraine, 6 with Belarus, and 5 with Kazakhstan (Hagemann, 1994). This is a substantial achievement considering the difficulty of negotiating with the former Soviet republics and the complexity of the coordination required within the U.S. government.

In arms-control verification, DNA conducts R&D activities supporting treaty monitoring of underground nuclear testing, strategic and intermediate-range nuclear force reductions, chemical weapons elimination, and conventional force reductions. DNA's efforts include the development and management of an on-site inspection technology testbed at Kirtland AFB, the development and transition to the On-Site Inspection Agency (OSIA) of hardware and techniques for inspections in treaty nations, and the development and maintenance of a Compliance Monitoring and Tracking System database to track treaty-accountable items.

3. How does DNA do it, and what are the costs? DNA has a small in-house staff dedicated to these activities, but most of the funds support contracts with private companies. In FY93, DNA spent about $337 million in this area,[7] with the vast bulk of this going for CTR activities. The funding allocated to this functional area is likely to increase significantly in FY94 and later years, particularly for CTR activities.

4. Observations and issues. The CTR activities in particular do not have a high RDT&E content but mostly are a coordination and procurement function. Although they require knowledge and management competence concerning nuclear weapons, these activities have little in common with DNA's historical specialized technical functions. They are a major part of the "redefinition" of DNA currently under way.

Conventional Defense Technologies

1. What is this functional area, and what are its future needs? DNA's charter allows it to "perform technical analyses, studies, and research on nonnuclear matters of critical importance to the Department of Defense where DNA has unique capabilities developed as part of its nuclear responsibilities" (U.S. DoD Directive, "Defense Nuclear Agency," 1993). This functional area is defined by the intersection of "nonnuclear matters of critical importance to [DoD]" with DNA's

---

[7]This includes activities in three DNA-designated areas: defense conversion program, cooperative threat reduction, and technical support for arms-control verification (DNA, 1994 Program).

"unique capabilities developed as part of its nuclear responsibilities." This definition encompasses a wide range of conventional weapons areas. The United States clearly has a continued need for research on a wide variety of conventional defense technologies to maintain military-technical advantages such as those demonstrated during Desert Storm.

DNA has been encouraged by OSD and by Congress (see the discussion in the appendix) to pursue technical opportunities in this area. Thus, the technical synergism between nuclear core competencies and some aspects of conventional munitions applications, when combined with similar overlaps in WMD applications areas, has provided the basis for recommendations within DoD that DNA become the Defense Special Weapons Agency or DSWA, to provide a DoD focal point for activities relating to all types of "special weapons"— chemical (offensive and defensive), biological (defensive only), and special-purpose conventional weapons, as well as nuclear weapons.

2. What does DNA do? Activities consistent with DNA's charter responsibility to "perform technical analyses, studies, and research on nonnuclear matters of critical importance to the Department of Defense where DNA has unique capabilities developed as part of its nuclear responsibilities" have steadily increased as a fraction of the DNA budget over the past few years and now cover a range of selected, conventional weapons areas.[8]

For example, using transport codes, DNA estimates collateral effects from conventional weapons strikes on WMD facilities. DNA also supports the development and employment of advanced conventional weapons (e.g., ballistic missile defense hit-to-kill systems, hard/buried structures munitions) by analyzing their lethality (using hydrodynamic codes) and supporting field tests (see Hagemann, 1994).

DNA has also developed bomb damage assessment (BDA) concepts (using gun camera data correlated with ground truth data from Desert Storm) that can be applied to hardened/buried structures. Its recent nonnuclear work on hardened structures and conventional (e.g., kinetic energy/high-explosive (HE) penetrators) and uncon-

---

[8]Some have referred to this as the "denuclearization of DNA."

ventional (e.g., penetrating radio frequency (RF) weapons) is a natural outgrowth of its earlier efforts in building nuclear-hardened buried structures and in understanding the effects of penetrating nuclear weapons on those structures and their mission equipment. In addition, DNA has applied its expertise in plasma physics to develop extended range 5-in. gun shells that do not exceed the safe operation pressures of the existing guns (see Hagemann, 1994).

Further, within the past year DNA has been directed by ATSD(AE) to assume the lead in developing a counterproliferation acquisition plan.[9]

3. How does DNA do it, and what are the costs? DNA out-sources almost all of its efforts in this area, relying on the contractor community and other government laboratories for the conduct of most of these activities.[10] RAND analysis of the DNA programs shows that DNA allocated about $147 million to this area in FY93.[11] This is currently a growth area for DNA, and the expenditure numbers for FY94 and later years are expected to be considerably higher.

4. Observations and issues. We do not question the value that these DNA activities have provided to the services, the Joint Staff, and other customers. However, we are concerned that the continued evolution of DNA in this nonnuclear, conventional- and special-weapons-technology direction must be *properly managed* to prevent a dilution and erosion in essential nuclear-related capabilities and expertise at DNA. The preservation of these capabilities remains a

---

[9]Thus far, this is a relatively small part of the overall DoD non/counterproliferation initiatives, but DNA's role could grow over time. For insights into the overall counterproliferation effort, see U.S. DoD, *Report on Nonproliferation,* 1994.

[10]The private sector companies and government laboratories involved in these activities are in most cases the same ones that have been involved in DNA's NWE research efforts over the years. This is only natural, since this is where the nuclear-related expertise that DNA is exploiting in this area actually resides.

[11]In its resource allocation and financial management system, DNA does not clearly separate these nonnuclear efforts from nuclear-related efforts. The activities RAND assigns to this category, judging by our analysis of the FY93 DNA budget and expenditures at the PDP level, are included in six DNA-designated areas: facility survivability, systems lethality, adaptive targeting, hard target interactions, counterproliferation technology, and computing and information systems (DNA, 1994 Program).

continuing requirement in the post-Cold War era. We will return to this issue in subsequent chapters.

DNA AS A WHOLE

What follows are a number of observations regarding DNA as a whole.

The Size of DNA

The four primary areas of DNA activity discussed above accounted for $907 million in expenditures in FY93. This number is in contrast to the FY93 DNA RDT&E appropriation (Program Element 62715H) of $399.7 million (see the appendix), the funding item that prompted the congressional request for the current study. What accounts for the difference?

As Table 3.1 shows, a number of items accounted for the difference in FY93:

- 6.3 funding from OSD RDT&E program elements for Verification Technology. (This funding is all used in the nuclear threat reduction and arms-control area discussed above.)

- RDT&E funds from the Ballistic Missile Defense Organization (BMDO) and other DoD agencies. (These funds are used in a variety of DNA program areas.)

- Operations and maintenance (O&M) funds. (These are used to provide operational and maintenance support to the DNA Headquarters and Field Command; to the Nuclear Weapons Stockpile Support activities discussed above, including the emergency response planning and exercise program; to Johnston Atoll; and to OSIA. (See Hagemann, *Course,* 1994.)

- Procurement Defense Agency (PDA) funds. (In FY94 these were primarily for Cooperative Threat Reduction (CTR)-related procurements. (See Hagemann, *Course,* 1994.)

- Military construction (MILCON) and real property maintenance funds.

- A variety of miscellaneous transfers.

Table 3.1

**DNA's Total FY93 Funding and Partial FY94 Funding
(in $ millions)**

	FY93	FY94
DNA technical base (PE 62715H)	399.8	235.0
6.3 verification technology	67.3	41.4
RDT&E subtotal from OSD	467.1	276.4
Ballistic Missile Defense Office RDT&E	42.1	8.5
Direct cite RDT&E	22.0	5.4
RDT&E total	531.2	290.3
Operations and maintenance	114.2	50.3
Procurement Defense Agency	237.8	?
Military construction	10.0	?
Real property maintenance	4.0	?
Miscellaneous transfers (net)	10.1	?
Total	907.3	TBD

Table 3.1 also shows a partial enumeration of DNA's FY94 funding. FY94 funding numbers for PDA, MILCON, real property maintenance, and miscellaneous transfers are not yet available. Of these, only PDA is likely to be significant. (Because of CTR activities, it could be even larger in FY94 than it was in FY93.) As these partial FY94 funding numbers indicate, DNA's various RDT&E funding accounts are undergoing major reductions, largely but not entirely as a result of the suspension of underground testing.

DNA Is Primarily an Executive Agency

As our discussions earlier in this chapter indicated, much of the effort in the various DNA functional areas is carried out by private sector companies and other government laboratories. Nuclear weapons stockpile support activities are, however, primarily provided by in-house personnel. About 90 percent of the total FY93 DNA funding was contracted to industry and government laboratories. Much of the expertise for which the United States may have a continuing need—i.e., core competencies in nuclear weapon effects and related technical areas—therefore resides in the research staffs of those private sector companies and government laboratories. DNA maintains a specialized core of technical managers to set priorities, establish research objectives, monitor the quality of the work, and provide continuity and corporate memory in the various DNA research areas.

Table 3.2 lists DNA's top ten contractors during FY93. The key government laboratories with DNA-developed expertise include the Air Force Phillips Laboratory (formerly known as the Space Technology Center), the Naval Research Laboratory, the Army Corps of Engineers Waterways Experimental Station, and the three DoE weapons laboratories.

DNA Is Changing

To gain a longer-term perspective on the changes that DNA is currently going through, it is useful to look at DNA funding profiles over the years. Based on data provided by the DNA Comptroller,[12] Figure 3.1 shows the annual appropriations of DNA and its predecessor agencies, the Armed Forces Special Weapons Project (AFSWP) and the Defense Atomic Support Agency (DASA), in constant FY93 dollars, from FY62 through FY94.

Table 3.2

**DNA's Top Ten Contractors, Based on FY93 DD350s
(in $ millions)**

Contractors	Obligations
Science Applications International Corp.	45.5
Scientific Ecology Group, Inc.	40.0
R&D Associates	29.7
JAYCOR	16.9
Mission Research Corp.	14.8
Maxwell Laboratories, Inc.	14.6
Cray Research, Inc.	14.1
Kaman Sciences Corp.	13.4
Physics International Company	12.1
EBASCO	10.2
Total	358.0

SOURCE: Hagemann, *Course,* 1994.

---

[12]The FY93 appropriation shown in Figure 3.1 varies from those shown in Table 3.1 by about $117 million. These funds were received from other agencies. Interagency funding transfer data were not available for all the years shown in Figure 3.1 and have not been included.

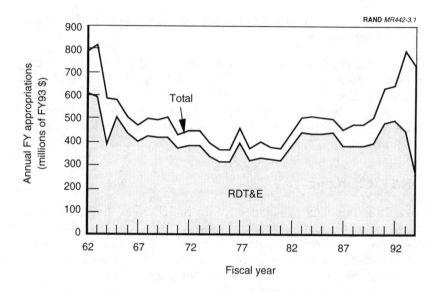

Figure 3.1—DNA Annual Appropriations Including CTR

The effect of the end of atmospheric testing in FY63 is apparent; from FY63 through FY76, DNA's RDT&E funding decreased by almost $300 million (in FY93 dollars of purchasing power). Starting in FY81, the RDT&E funding grew approximately $190 million, largely as a result of the Strategic Modernization Program. With FY92 came a ban on underground testing and the RDT&E funding for FY93 and FY94 experienced large declines. Until FY91, the total DNA funding has followed the RDT&E funding levels and remained at a fairly constant increment above it. DNA was primarily an RDT&E organization.

The divergence between total DNA funding and RDT&E funding in the last few years is due to the CTR funding, which is increasing rapidly. CTR funding was $12 million in FY92 and should be $400 million or even higher in FY94. CTR funding largely goes to procure equipment used in dismantling nuclear weapons and transporting, storing, or securing sensitive components, not for RDT&E activities. Although only 43 DNA personnel are currently directly involved with CTR, the workload on the organization's infrastructure is clearly growing. The Comptroller, the Acquisition Management Organization, and the DNA leadership are bound to be affected. DNA is currently experiencing a cultural change and is no longer an agency dominated solely by RDT&E but is becoming one with an equally large procurement dimension.

Chapter Four

THE CONGRESSIONAL OPTIONS

INTRODUCTION

In this chapter we address and assess options for carrying out the functions currently performed by DNA suggested in the conference language leading to this study. The congressional language proposed the following options:

Option 1. Transferring DNA's functions to the individual armed services and ARPA under an arrangement whereby a service or ARPA becomes the executive agent for the entire DoD for the function or functions transferred.

Option 2. Maintaining DNA as a separate agency under the plan proposed by the DoD in a letter to the congressional defense committees on June 25, 1993, to adapt the agency to the conditions of the new international security environment.

Option 3. Transferring DNA's functions to the Department of Energy nuclear weapons laboratories.

Option 4. Combining any of the previously listed options.

Option 5. Reorganizing DNA to significantly reduce the agency's operating, management, administrative, and other overhead costs.

Congress further suggested that the following issues figure in the evaluation of these options:

1. Any effect on the federal government's ability to maintain critical expertise in nuclear and conventional defense areas and to perform effectively functions now performed by DNA.

2. Annual cost savings.

3. A reasonable implementation schedule.

We begin this chapter by describing our interpretation of each congressional option. Next we discuss the assessment criteria we used and make some observations regarding the cost implications of the various congressional options. We then consider each of the four current DNA functional areas outlined in the previous chapter in turn, applying our criteria and assessing the various congressional alternatives. In this process, we identify a number of pros and cons associated with each option, as well as risks that are involved and ways of mitigating those risks. This leads us to an identification of optional recipient organizations for each of the principal DNA functional areas if a decision were made to distribute some or all of DNA's functions. We also assess those congressional options providing for the continued existence of DNA. This chapter ends with a summary of the conclusions that we draw from our assessment analysis.

THE FIVE CONGRESSIONAL OPTIONS: WHAT DO THEY INVOLVE?

In carrying out our analysis, we have interpreted the five congressional options as follows:

Option 1: Transfer DNA's functions to the armed services or ARPA. As stated, this option would involve the transfer of one or more of the four principal groups of DNA functions[1] either to one or more of the military services or to ARPA. For reasons that will become apparent in the discussions to follow, we have broadened the scope of this option to include transfers to other military organizations, e.g., joint commands such as STRATCOM or tri-service offices, and to other

---

[1]As discussed in Chapter Three, these four functional groups are nuclear weapons stockpile support, nuclear weapons effects research and operational support, nuclear threat reduction and arms control, and conventional defense technologies.

defense agencies, e.g., OSIA. Within the services, likely homes would depend on the nature of the DNA function in question. For example, for research-oriented functions, one or more of the service laboratories are obvious candidates.

In any such transfer, whether to one of the military services, STRATCOM, ARPA, or OSIA, what is involved here would be the transfer of DNA's current executive responsibilities and authorities for the functional area in question, along with associated funding, and any provider-of-services role that DNA currently plays.[2]

Option 2: Maintain DNA as a separate agency, allowed to adapt to the conditions of the new international security environment. This option is often termed the "DSWA option," referring to the proposal currently circulating in DoD circles that DNA evolve into a Defense Special Weapons Agency or DSWA, to provide a DoD focal point for activities relating to all types of "special weapons"—chemical (offensive and defensive), biological (defensive only), and special-purpose conventional weapons, as well as nuclear weapons. We have interpreted this option in precisely that fashion.

Option 3: Transferring DNA's functions to the DoE nuclear weapons laboratories. This option involves the transfer of one or more of the four principal groups of DNA functions to one or more of the DoE weapons laboratories: Los Alamos National Laboratory (LANL), Lawrence Livermore National Laboratory (LLNL), and Sandia National Laboratory (SNL). As in Option 1, any such transfer would certainly include DNA's current executive responsibilities and authorities for the functional area in question, along with associated funding. In addition, this option could also involve the transfer of the provider-of-service role currently performed by DNA contractors to the DoE laboratory in question.

Option 4: Combinations of any of these options. This option involves combinations of Options 1 and 3, with one set of DNA functions going to one organization and another set to a different organization, putting each set of functions in the best possible place.

---

[2]As indicated in Chapter Three, DNA currently plays a provider-of-services role only in the nuclear weapons stockpile support area. In the other functional areas, DNA is primarily an executive agency.

Option 5: Reorganizing DNA to significantly reduce the agency's operating, management, administrative, and other overhead costs. This option is often termed the "lean DNA" option. We have assumed that it would involve a continuation of DNA activities in all (or almost all) of its current functional areas, with a core group of DNA technical managers and project officers who manage the research efforts of private contractors and other government research organizations external to DNA. Consistent with the letter of this option, we have not included an assessment of whether and to what extent substantive functions of DNA could be reduced in view of the changing national security environment.

ASSESSMENT CRITERIA

Thus, we understand from the congressional language requesting this study that the objective is to maintain the main essential functions performed by DNA while significantly reducing the cost to the U.S. taxpayer. This suggests two primary assessment criteria:

- The *risks,* if any, to the maintenance of a function associated with the adoption of a particular option.

- The *cost effect,* positive or negative, associated with the adoption of a particular option.

In this section we discuss our handling of assessment criteria dealing with *risk.* The subject of *costs* is treated in the next section.

As discussed in Chapters Two and Three, DNA currently performs three main functions for which there is an enduring need, at some level of effort, in the post-Cold War era:

- Nuclear weapons stockpile support, which contributes to the maintenance of the U.S. nuclear stockpile.

- Nuclear weapons effects research and operational support, which contribute to the maintenance of an ability to employ nuclear weapons effectively and to conduct nuclear operations in a nuclear environment if either of these should become necessary in the future.

- Nuclear threat reduction and arms control, which contribute to the maintenance of a capability to pursue effective nuclear (and other WMD) threat reduction and arms-control activities.

The nature of the principal risks to the maintenance of these functions associated with the various congressional options will differ for each functional area. Later in this section we will discuss each area in turn and identify what we consider to be the principal risks. Whatever the nature of the risks, some metric is needed to quantify them.[3] In attempting to do this, we have adopted three designators describing the relative magnitudes of the various risks:

- **Minimal risk:** One that is unlikely to threaten the essential performance of a function. These risks can essentially be ignored because they are insignificant.

- **Manageable risk:** One that could threaten the performance of a function but that could be effectively mitigated in a feasible fashion that we have been able to identify.

- **Substantial risk:** One that could seriously threaten the performance of a function and that cannot be effectively mitigated in any feasible fashion that we can identify.

As defined, these risk designators involve assessments of the effect of a given risk on the function in question, as well as assessments of the efficacy of mitigation measures. The manner in which we apply these designators will become clear in subsequent discussions.

COST CONSIDERATIONS

As discussed in Chapter Three, DNA is a combination of two different types of organizations. One, the "provider DNA" carries out the nuclear weapons stockpile support activities, conducting its efforts using in-house personnel, assigned primarily to the DNA Field Command, with a small number at DNA Headquarters. In FY93, the provider DNA had about 250 personnel (140 military and 110 civil-

---

[3]As the reader will no doubt conclude from subsequent discussions, "quantify" is probably much too strong a term. At best, we have attempted to "qualify" the relative magnitudes of the various risks.

ian)[4] and consumed about $21 million (which includes military personnel costs of approximately $11 million)[5] in performing its functions.

The second, or "executive DNA," works in the other three principal DNA functional areas: nuclear weapons effects research and operational support, nuclear threat reduction and arms control, and conventional defense technologies. DNA funds others for almost all of its efforts in these three areas, relying on the contractor community and other government laboratories for the conduct of most of the activities. In FY93, the executive DNA had about 1100 personnel (350 military and 750 civilian) and consumed about $920 million (including military personnel costs) in performing its functions.

DNA has identified a series of internal restructuring actions involving the streamlining and modernizing of procedures and operations. DNA management estimates that these actions could lead to a reduction of over 250 personnel spaces over a five-year period (Hagemann, Course, 1994).[6] These internal restructuring actions would also save about $18 million per year in O&M and military personnel costs.[7]

We have also considered the overall magnitude of the cost savings that might result from one or another of the congressional options that involve the disestablishment of DNA and the dispersal of its

---

[4]These numbers are approximate and are based on data contained in (Hagemann, Course, 1994) and (DNA, Field Command, RAND, 1994).

[5]The personnel costs of the military personnel assigned to DNA are not paid for out of the DNA O&M budget, but rather are charged to the individual services. In FY93, DNA had 498 authorized military personnel, about two-thirds officers and one-third enlisted. Assuming about 95 percent of these spaces were filled and using average personnel cost figures of $83,000 per year per officer and $53,000 per year per enlisted, the total DNA military personnel costs in FY93 were about $35 million. The corresponding FY94 figures are 439 authorized military personnel and total DNA military personnel costs of about $31 million. These military personnel costs were not included in the DNA functional area expenditures listed in Chapter Three.

[6]It may be possible to carry out such a reduction over a shorter period or with deeper personnel cuts. We were not able to address these possibilities in the study.

[7]Only about two-thirds of this savings would show up in the DNA budget, since the military personnel costs are accounted for elsewhere in the DoD budget.

functions to other organizations.   Our conclusion is that the cost savings will not be large, *provided that the functions transferred continue to be carried out at about the same levels of activity and effectiveness in their recipient organizations as they were in DNA.*

This is a key assumption.   The same levels of activity imply roughly the same number of personnel working on a particular activity.   The same levels of effectiveness imply roughly the same caliber of people, in terms of education and experience, and therefore roughly the same salary levels.   Similar numbers of people and similar salary levels imply roughly similar direct personnel costs and employee benefit costs.   This leaves two possible areas of savings if DNA is disestablished and its functions dispersed: (1) lower indirect support costs could be realized by absorbing overhead functions into already existing organizational structures, and (2) some or all of the top management personnel of DNA could be replaced by the top management personnel of the recipient organizations.

Regarding support costs, possible areas of cost savings include rent and utilities, as well as human resources, security, contracting, and comptroller functions.   Table 4.1 lists DNA's expenditures in these areas.   In FY93 they totaled about $10 million.[8]   It also lists the personnel costs associated with the top 15 DNA management positions.   In FY93 these were about $4 million.

Assuming that the agencies receiving the DNA functions can carry them out without increased costs for rent or utilities, and without having to augment their existing human resources, security, contracting, and comptroller staffs or their executive management personnel, the total annual savings would be about $19 million: the $13.6 million shown in Table 4.1 plus roughly $5 million in military personnel costs.[9]

---

[8]These numbers do not include the military personnel costs associated with the performance of these functions. We estimate these to be about $5 million.

[9]DNA claims that FY93 costs are no longer a valid benchmark, because of various cost reductions that have already taken place. In any event, the maximum annual savings that can be realized in this way are probably in the neighborhood of $20 million.

Table 4.1

**Selected DNA FY93 Expenditure Categories
(in $ millions)**

Category	Expenditures
Rent	2.19
Utilities	.82
Human resources	1.46
Security	1.11
Contracts	2.19
Comptroller	1.81
Subtotal	9.58
Executive salaries and benefits	4.00
Total	13.58

SOURCE: DNA Comptroller's Office.

Substantial cost savings require substantial reductions in activity levels. If this is done, the cost savings will be about the same, whether the functions remain at DNA or are transferred elsewhere. So if significant cost saving is the objective, transferring functions out of DNA is not the answer; reducing the level of activity on those functions is the answer, assuming that such reductions are warranted. Although reduced activities might occur as a result of transferring DNA functions to other organizations, which in turn assign lower priority to them, this is not a sound way to determine how much funding each function merits, especially given the importance of nuclear weapons functions.

EVALUATION OF OPTIONS

With all of the above as a necessary preamble, we now turn to an assessment of the congressional options. For this purpose, we have divided the options into two categories:

- Those in which DNA is disestablished and its functions are all transferred (Options 1, 3, and 4).

- Those in which DNA continues to exist and perform most or all of its current functions (Options 2 and 5).

OPTIONS IMPLYING THE DISESTABLISHMENT OF DNA

Options 1 and 3

For these options (as well as Option 4, which will be discussed later), the dominant considerations will be:

- The risk associated with transferring a given DNA function to a given organization.

- The likely effectiveness and efficiency with which a given organization can perform a given DNA function.

The principal risks to be considered will differ for each functional area: nuclear weapons stockpile support, nuclear weapons effects research and operational support, nuclear threat reduction and arms control, and conventional defense technologies. We now discuss each of these in turn.

Nuclear Weapon Stockpile Support. As discussed in Chapter Three, DNA's primary role here is in the management, conduct, and auditing of stockpile support activities to ensure the safety, security, reliability, and, to a much more limited extent in the future than during the Cold War, the survivability of that portion of the warhead stockpile under DoD custody. DNA conducts these efforts using largely in-house (primarily service) personnel, assigned to the DNA Field Command. This is clearly an area for which the United States will have an enduring need in the post-Cold War world, albeit at a reduced scale of activity as the U.S. nuclear stockpile declines in size.

In assessing the possible transfer of this DNA function to other organizations, three factors are of particular concern:

- The ability of the recipient organization to maintain an "independent" audit of the stockpile support activities of other DoD components. This is important to guarantee the adequacy of activities ensuring the safety, security, and reliability of the nuclear stockpile, in an era of ever tightening budgets, which could provide motivations to cut corners.

- The accountability of the recipient organization to the Secretary of Defense, through some straightforward management channel.[10] The Secretary of Defense is charged by law with responsibility for the safety and security of the nuclear weapon stockpile under DoD custody. Whoever acts as the secretary's agent in fulfilling this responsibility must be accountable in a straightforward fashion, with a minimum of intervening management or command levels and without raising the specter of a loss of civilian control of nuclear weapons.

- An institutional "culture" compatible with the stockpile support and auditing activities. This factor recognizes that the corporate culture of an institution manifests itself in many ways—the nature of its primary mission, the organizational means established for achieving its goals, and the characteristics of the staff responsible for the mission. This institutional culture must be compatible with the functional demands of the stockpile support activity.

In Table 4.2 we identify a number of pros and cons associated with the transfer of this functional area to a particular recipient organization (per Options 1 and 3).[11]

It appears from this enumeration of pros and cons that the most promising receiving organizations are:

- For Option 1 (services/ARPA): a tri-service office or STRATCOM.

- For Option 3 (DoE labs): no fully satisfying choices here.

Table 4.3 lists the principal risks associated with each of these choices and the available mitigation approaches and gives our overall risk assessment for each transfer option, assuming mitigation steps are taken.

---

[10]The DNA management channel for this functional area is through the ATSD(AE) for coordination and DDR&E for budget authority.

[11]In this table "tri-service office" denotes a jointly staffed office reporting directly to OSD. (The stockpile support portion of DNA Field Command could form the cadre for such an office.)

This analysis shows that the preferred receiving organization for DNA's nuclear weapon stockpile support functions would be either a tri-service office reporting to OSD (first choice) or STRATCOM (second choice). The risks are probably fewer with the tri-service office, but the risks with STRATCOM seem manageable, and the choice of STRATCOM could lead to greater organizational coherence and efficiency given that CINCSTRAT will have essentially all of the operational U.S. nuclear weapons under his purview.

Nuclear Weapons Effects Research and Operational Support. As discussed in Chapter Three, DNA's primary role here has been as the DoD center of excellence for understanding and documenting the effects of nuclear weapons on systems, for both vulnerability/hardening and targeting support to the services and CINCs. DNA has provided the detailed understanding of NWE required to support the effective employment of nuclear weapons by U.S. military forces and the conduct of military operations by those forces in a nuclear (or potentially nuclear) environment. DNA contracts out almost all of its efforts in this area, relying on the contractor community and other government laboratories for the conduct of most of these activities. DNA maintains a core of technical managers to set priorities, establish research objectives, monitor the quality of the work, and provide continuity and corporate memory.

In assessing the possible transfer of this DNA function to other organizations, three factors are of particular concern:

- The institutional culture and technical capabilities of the recipient organization. As in the case of Stockpile Support, the institutional culture of that organization must be compatible with the functional demands of the NWE research and operational support activity. These demands will frequently involve complicated technical issues, with which the recipient organization must be comfortable.

- The recipient organization's understanding of the needs and perspectives of the military customer base that this function supports. This customer base includes operators and designers of weapon systems carrying nuclear weapons or possibly operating in a nuclear weapon environment. It also includes the weapon design community and those involved in the safety, handling,

Table 4.2

Pros and Cons Associated with the Transfer of DNA's Nuclear Weapon Stockpile Support Functions

Receiving Organization	Pros	Cons
Option 1		
Air Force and/or Navy	Selected Air Force and Navy personnel should have a continued interest in nuclear matters	Possible conflict of interest in performing the stockpile audit function, since the Air Force and Navy will continue to operate nuclear weapon systems[a]
Army	No conflict of interest	Probable lack of service interest, since the Army has given up all of its nuclear weapons
Tri-service office (reporting to OSD)	Ease of implementation (e.g., turn DNA Field Command into a tri-service office reporting directly to OSD)	Could require creation of a new organization
	No conflict of interest	
	Straightforward reporting chain to the Secretary of Defense	
STRATCOM	Ease of implementation (e.g., have ex-DNA Field Command report directly to CINCSTRAT)	Possible conflict of interest, unless stockpile audit agency reports directly to CINCSTRAT, completely separate from nuclear delivery systems' chain of command
	Organizational coherence and efficiency (STRAT-COM will control most of the remaining U.S. nuclear weapons)	May require special reporting chain to the Secretary of Defense
	Obvious motivation regarding nuclear matters	

Table 4.2 (continued)

Receiving Organization	Pros	Cons
ARPA	No conflict of interest	Mismatched institutional culture. ARPA is primarily a high-tech R&D organization that frequently shifts its research focus in response to technological opportunities. The DNA stockpile support activity is a low-tech operational mission with continuing responsibilities for the indefinite future
	Straightforward reporting chain to the Secretary of Defense can be established	
Option 3 DoE weapon laboratories	Have all required technical capabilities	Mismatched institutional culture. DoE labs are primarily RDT&E organizations. This is primarily an operational mission
	Already providing some specialized technical aspects of stockpile support	Possible conflict of interest, insofar as the auditing function is concerned, since each DoE lab has a vested interest in various stockpile weapons
		Circuitous reporting chain to the Secretary of Defense

[a]The history here is mixed. The Navy has instituted an independent audit of the nuclear weapons in Navy custody, with the auditing personnel reporting to a chain of command independent of the chain of command controlling the weapons, all the way up to the senior Navy leadership. Judging from historical data, this Navy independent audit function appears to perform just as well as the fully independent DNA audits. Existing Air Force audit activities are less independent of the weapon system chain of command and have a more mixed performance history.

Table 4.3

Risk Assessment: Transferring the DNA Nuclear Weapon Stockpile Support Functions

Transfer Option	Principal Risks	Mitigation Measures	Risk Assessment
Option 1 Tri-service office	No significant risk	None required	Minimal
STRATCOM	Maintenance of audit independence	Have audit agency (ex-DNA Field Command) report directly to CINCSTRAT, completely separate from nuclear delivery systems' chains of command	Manageable
	Accountability to the Secretary of Defense	Make CINCSTRAT directly responsible to the Secretary of Defense for stockpile assurance (not through the JCS)	
Option 3 DoE weapon laboratories	Mismatched institutional culture	No obvious, straightforward measures likely to be adequately effective	Substantial
	Accountability to the Secretary of Defense		
	Possible conflict of interest		

and caring for the nuclear weapon stockpile as well as those who are charged with planning for potential future situations in which nuclear weapons may be involved. Of necessity, the NWE RDT&E agenda should reflect the understanding of these customer viewpoints and be responsive to changes.

• The recipient organization's ability to provide appropriately effective proponency regarding NWE matters. By "effective proponency" we mean the continuing need for the organization responsible for this function to be able to effectively articulate and defend the RDT&E agenda for this function based on operational contingencies that the customer base believes are important as well as the ability to express legitimate concerns about the survivability requirements for current and future weapon systems.   By "appropriate" we mean effective proponency guided by the proper priorities: i.e., priorities appropriate to the role nuclear weapons will play in the post-Cold War era, with not too much emphasis on nuclear matters, but not too little either.

Table 4.4 identifies a number of pros and cons associated with the transfer of this functional area to a particular recipient organization (per Options 1 and 3).

From this enumeration of pros and cons, it appears that the most promising receiving organizations are:

- For Option 1 (services/ARPA):   One of the leading service laboratories, preferably one with a past history of research excellence in nuclear-related matters.   (The Air Force Phillips Laboratory is probably the leading contender.)

- For Option 3 (DoE labs): One or more of the three DoE weapon laboratories.

Table 4.5 lists the principal risks associated with each of these choices and the available mitigation approaches and gives our overall risk assessment for each transfer option based on the application of these mitigation approaches.

This analysis shows that the preferred receiving organization for DNA's nuclear weapons effects research and operational support functions appears to be one or more of the DoE weapon laboratories.

Nuclear Threat Reduction and Arms Control. As discussed in Chapter Three, DNA's role in this area includes: (1) a variety of activities supporting ongoing U.S. nuclear threat reduction efforts, including most prominently those associated with the CTR program and (2) R&D activities supporting compliance monitoring of a number of arms-control treaties. These areas are of ever increasing importance as the United States gets deeper into the post-Cold War era and as DNA's activities (particularly those associated with CTR) are in the process of considerable expansion.

Table 4.4

Pros and Cons Associated with the Transfer of DNA's Nuclear Weapons Effects Research and Operational Support Functions

Receiving Organization	Pros	Cons
Option 1		
Selected Army, Navy or Air Force laboratories	Intimate understanding of customer base	Services exhibiting increasing disinterest in nuclear matters
	Leading service laboratories have well-matched institutional culture	Leading service laboratories have largely gotten out of nuclear-related research[a]
ARPA	Used to dealing with complicated technical issues	Apparent lack of current interest in most nuclear matters
		Institutional culture favors focused research activities leading to near- or mid-term payoffs, not long-term knowledge and capability maintenance efforts
Option 3		
DoE weapon laboratories	Have all technical capabilities required for NWE research	DoE labs may not be as closely coupled to the military customer base as they once were
	Well-matched institutional culture	Nuclear-weapon-related research at the nuclear weapon laboratories is undergoing a steep decline[b]
	Proven track record of effective proponency on nuclear matters	

[a]In the 1960s, 1970s, and early 1980s, the Air Force Weapons Laboratory was also a center of excellence in NWE research, rivaling DNA in some areas. At that time, it could have taken over many if not most of the DNA NWE research functions if necessary. In recent years, it has been reoriented (and renamed) as part of the Air Force Phillips Laboratory and has largely refocused its efforts away from nuclear research areas to concentrate on space science and technology.

[b](1) Lawrence Livermore National Laboratory (1993), (2) Los Alamos National Laboratory (1993), (3) Sandia National Laboratory (1993).

Table 4.5

Risk Assessment: Transferring the DNA Nuclear Weapons Effects Research and Operational Support Functions

Transfer Option	Principal Risks	Mitigation Measures	Risk Assessment
Option 1 A leading service laboratory	Service and laboratory disinterest in getting back into nuclear-related research	High-level OSD and service leadership insistence that selected laboratory get back into nuclear research and do a good job, and continued management emphasis on this	Uncertain. Could be manageable or substantial
Option 3 One or more DoE weapon laboratories	Inadequate coupling of DoE labs to military customer base	Increased liaison activities, including exchange of personnel	Manageable
	Nuclear weapons research becoming subordinate to other interests	Select DoE labs whose management exhibits clear desire to stay in nuclear weapon effects business	

DNA has a small in-house staff dedicated to these areas, but most of the funds are contracted to private companies, with procurement of relatively routine equipment for the CTR program accounting for a major fraction of the activities.

In assessing the possible transfer of this DNA function to other organizations, two factors are of particular concern:

- The institutional culture of the recipient organization, which must be compatible with the functional demands of the nuclear threat reduction and arms-control activity, particular the CTR portion, which is primarily a coordination and procurement function.

- The accountability of the recipient organization to the ATSD(AE), and to other OSD offices charged with arms-control responsibilities, through some straightforward management channel.[12]

Table 4.6 identifies a number of pros and cons associated with the transfer of this functional area to a particular recipient organization (per Options 1 and 3).[13]

It appears from this enumeration of pros and cons that the most promising receiving organizations for this functional area are:

- For Option 1 (services/ARPA): OSIA for the nuclear threat reduction coordination and procurement function.

- For Option 3 (DoE labs): One of the DoE weapon laboratories, but just for the arms-control verification technology portion of the activity, not the nuclear threat reduction coordination and procurement function.

Table 4.7 lists the principal risks associated with each of these choices and the available mitigation approaches and gives our overall risk assessment for each transfer option based on the application of these mitigation approaches.

From this analysis, the preferred receiving organizations for DNA's nuclear threat reduction and arms-control functions appear to be either OSIA (augmented by a cadre of DNA R&D technical managers) for the entire functional area or a split assignment, with OSIA taking over the nuclear threat reduction activities and one of the DoE weapon laboratories taking over the arms-control verification technology R&D activities. If the R&D management capabilities of OSIA can be adequately enhanced, the OSIA option may be preferable, since it could create an even closer coupling between the on-site inspection operations currently conducted by OSIA and the verification technology R&D activities supporting those operations.

---

[12]CTR is a legislated responsibility of OSD. The Secretary of Defense has delegated the responsibility for CTR implementation to the ATSD(AE).

[13]As noted above, we have broadened the scope of Option 1 (services/ARPA) to include other defense agencies such as OSIA.

Table 4.6

Pros and Cons Associated with the Transfer of DNA's Nuclear Threat Reduction and Arms-Control Functions

Receiving Organization	Pros	Cons
Option 1		
One or more military services	Institutional culture compatible with complex coordination and procurement activities	May require a special reporting chain to OSD
ARPA	Straightforward reporting chain to OSD could be established	Mismatched institutional culture. ARPA is primarily a high-tech R&D organization, ill-suited to large-scale procurement of relatively routine equipment
OSIA	Direct reporting chain to OSD Capable of supporting complex coordination and procurement activities Could create even closer coupling between on-site inspection operations and verification technology R&D activities	May be weak in technical capabilities required to manage arms control verification technology R&D May be weak in contract management and related services required for CTR procurements
Option 3		
DoE weapon laboratories	Have all required technical capabilities for arms-control verification technologies portion of this functional area	Somewhat mismatched institutional culture. The DoE labs are primarily high-tech RDT&E organizations. However, they have engaged in large-scale procurement activities in the past when required by their mission Special reporting chain to OSD required

Table 4.7

Risk Assessment: Transferring the DNA Nuclear Threat Reduction and Arms-Control Functions

Transfer Option	Principal Risks	Mitigation Measures	Risk Assessment
Option 1			
OSIA	Ability to adequately manage arms-control verification technology R&D	Transfer selected DNA personnel to OSIA to serve as R&D management cadre	Manageable
	Ability to adequately manage CTR procurements	Transfer selected DNA contractual and legal personnel to OSIA	
Option 3			
DoE weapon laboratory	Somewhat mismatched institutional culture, but one of the labs could do it if they really wanted to	Split the functions. Give the arms-control verification technology R&D to a DoE lab.	Manageable
		Give the nuclear threat reduction coordination and procurement function to someone else (i.e., OSIA)	
	Accountability to OSD for CTR activities	Establish special reporting chain, or split the functions	

Conventional Defense Technologies. As discussed in Chapter Three, this area represents an extension of DNA's activities into a number of different nonnuclear military technology areas—areas in which DNA can exploit capabilities originally developed as part of its nuclear responsibilities. DNA contracts with others for almost all of its efforts in this area, relying on the contractor community and other government laboratories for the conduct of most of these activities.

The United States clearly has a continued need for research on a wide variety of conventional defense technologies, including the areas being exploited by DNA. The military services have active R&D programs in many/most of these same areas and have access to the same contractor base and government laboratories on which DNA relies for most of its expertise. This makes the transfer of activities in this functional area to other agencies especially straightforward.

Furthermore, there is no compelling reason why DNA's current research activities in this area must be kept together as a group if they are transferred. Rather, they could be divided based on the particular technologies involved or on the particular area of military operations being supported.

In assessing the possible transfer of activities in this DNA functional area to other organizations, three factors are of particular concern:

- The institutional compatibility of the recipient organization, with both the technology area being exploited and the military application area being targeted in the RDT&E activity in question.

- The recipient organization's understanding of the needs and perspectives of the operational military users that the particular RDT&E activity supports.

- The recipient organization's involvement in the requirements determination and resource allocation process affecting the military R&D area in question.

Table 4.8 identifies a number of pros and cons associated with the transfer of this functional area to a particular recipient organization (per Options 1 and 3).

It is obvious from this enumeration of pros and cons that the various service R&D organizations and ARPA would be the national receiving organizations for this functional area. Table 4.9 reinforces this by listing the risks associated with this choice and giving our overall risk assessment.

This analysis shows that the preferred receiving organizations for DNA's conventional defense technologies function are clearly the service R&D organizations and ARPA. The only real risk in such transfers is that some DNA research activities could possibly fall by the wayside. If the activities are clearly valuable, however, and properly "advertised" to likely recipient organizations, this risk should be minimized.[14]

---

[14]One can also take the position that if a specific DNA research activity is not picked up by one of the service R&D organizations or ARPA, it may not really be that valuable.

Table 4.8

Pros and Cons Associated with the Transfer of DNA's Conventional Defense Technologies Function

Receiving Organization	Pros	Cons
Option 1		
Various service R&D organizations	Compatible institutional culture	No obvious cons
	Intimate understanding of operational needs and perspectives	
	Participation in RDT&E requirements determination and resource allocation process	
ARPA	Compatible institutional culture	May be detached from operational needs and perspectives in some areas
	Generally adequate understanding of operational needs and perspectives	
	Participation in RDT&E requirements determination and resource allocation process	
Option 3		
DoE weapon laboratories	Compatible institutional culture	Detached from operational needs and perspectives in many areas
		Disconnected from military RDT&E requirements determination and resource allocation process

Option 4

Our method for identifying and assessing the best option for dismantling DNA and distributing its functions among several recipient organizations is to combine the lowest-risk transfer by function from Options 1 and 3. Table 4.10 assesses this approach.

Table 4.9

**Risk Assessment: Transferring the DNA Conventional Defense
Technologies Function**

Transfer Option	Principal Risks	Mitigation Measures	Risk Assessment
Option 1			
Various service R&D organizations	No obvious risks, as long as individual R&D organizations acquire research activities only in their area of specialization	None required	Minimal
ARPA	No obvious risks, as long as ARPA acquires research activities compatible only with its overall research agenda	None required	Minimal

Table 4.10

Option 4 Summary

Function	Receiving Organization	Risks
Stockpile support	Tri-service office reporting to OSD	Minimal
Nuclear weapons effects	DoE labs	Manageable. Would require increased DoD-DoE linkage and selection of labs with clear nuclear weapons commitment
Nuclear threat reduction and arms control	OSIA (nuclear threat reduction)	Manageable
	DoE labs (arms-control verification R&D)	Manageable
Conventional defense technologies	Services or ARPA	Minimal

Thus, when viewed function by function, we see minimal to manageable risk in dismantling DNA and distributing its functions to those organizations best suited to receive and manage them. However, we have several larger concerns that are especially acute in regard to this "least-risk" combination:

- It could aggravate the overall problem of fragmentation and the associated risks (lack of critical mass, reduced sense of priority, strained interdependencies).

- The benefits (minus cost savings) would not appear to outweigh even minimal-manageable risks, especially if there is any chance that the risks have been underestimated.

- Potentially important options concerning a larger consolidation of the U.S. nuclear weapons infrastructure (as presented in the next chapter) could be impeded if this particular option is chosen.

OPTIONS IMPLYING THE CONTINUED EXISTENCE OF DNA

We turn next to the two congressional options in which DNA continues to exist and continues to perform most or all of its current functions, Options 2 and 5.

"Lean DNA" (Option 5)

We begin with Option 5, often termed the "lean DNA" option. In this option, DNA would be reorganized to significantly reduce the agency's operating, management, administrative, and other overhead costs. DNA would continue activities in all (or almost all) of its current functional areas, with a core group of DNA technical managers and project officers who manage the research efforts of private contractors and other government research organizations external to DNA.

Earlier in this chapter we made a number of observations regarding cost issues associated with the current DNA program and with various possible changes in that program. Based on information provided by DNA and our own analysis, it appears that annual cost savings of somewhat less than $20 million can be achieved by a near-term "leaning" of the DNA structure, involving a series of streamlining actions and the reduction of over 250 civilian and military personnel spaces. Greater cost savings would obviously require more dramatic changes in the structure of DNA. In the ultimate, "ARPA" limit, additional cost savings could be realized, about two-thirds of which would show up on the DNA budget.[15] However, assuming that the overall scale of DNA activities remained roughly the same as it is today, this would likely represent a significant but still small savings compared with DNA's total expenditures in FY93 of $907 million.

Defense Special Weapons Agency (Option 2)

This option envisages the maintenance of DNA as a separate agency, allowed to adapt to the conditions of the new international security environment. As mentioned above, this option is often termed the "DSWA option," referring to the proposal currently circulating in DoD circles that DNA evolve into a Defense Special Weapons Agency, to provide a DoD focal point for activities relating to all types of "special weapons"—chemical (offensive and defensive), biological (defensive only), and special-purpose conventional weapons, as well as nuclear weapons.

Indeed, the DNA senior leadership has expressed serious concern, given the ongoing reduction in DoD budgets for all nuclear-related activities, that DNA or any other agency focused solely or primarily on nuclear activities can long remain viable. The workforce would gradually shrink as budgets continue to decline, the best people would begin to leave, the best and the brightest of the new generations would avoid the agency, the challenge and excitement would gradually seep out of its research agenda, and its vitality would gradually ebb away.

Recognizing this risk, this option would allow the diversification into new and challenging areas that we have identified as the most straightforward way of mitigating the risk of atrophy faced, in the post-Cold War world, by an agency focused solely on nuclear matters.[16] Given the challenges facing the United States in the special

---

[15]As mentioned above, the cost of military personnel assigned to DNA, currently about 30 percent of the DNA staff, does not show up on the DNA budget but rather is charged to the individual services.

[16]We are not the first to think of this. Almost every research organization that played a prominent role in the U.S. nuclear infrastructure during the Cold War is actively pursuing diversification, for much the same reasons.

weapons area—preventing the proliferation of WMD, countering WMD proliferation if and when it has occurred, etc.—we recognize that allowing DNA to evolve in this direction will provide the challenge necessary to maintain its vitality and to attract, retain, and motivate good people.

The United States clearly has a continued need for research on a wide variety of special weapon technologies, to maintain for use in future conflicts (if and when they occur) the military-technical advantages that the United States demonstrated during Desert Storm, and to deal with future WMD proliferation. DNA and its contractor base have a number of specialized capabilities developed as part of their nuclear research activities that can have synergistic application in a variety of special weapons areas. Furthermore, the existence of a high-quality, center-of-excellence research agency to serve as a focal point for U.S. technical activities across the spectrum of special weapons should lead to greater coherence in dealing with WMD-related issues.

At the same time, we have been concerned about precisely the opposite risk: that the continued shift of DNA in this nonnuclear, special-weapons-technology direction could dilute and erode essential nuclear-related capabilities and expertise at DNA, whose preservation remains a continuing requirement in the post-Cold War era. The most capable DNA technical managers and staff members may be guided away from supporting their nuclear activities by the attractiveness of the new special weapons research projects, which would be perceived as the areas of personal and funding growth. Such a perception, and the associated risks, would be accentuated if DoD and DNA management showed greater attention to and interest in DNA's nonnuclear initiatives. Therefore, there is a point of view that, to have high confidence of maintaining essential nuclear capabilities, DNA should focus exclusively on nuclear matters and not be diverted into nonnuclear areas.

It is probably a question of leadership direction, commitment, and vigilance that would determine whether DNA, if it evolves into an agency doing work in the new special weapons areas, would still maintain the essential nuclear capabilities. Without such leadership, the essential nuclear capabilities could very well erode during the transition. The risk is real but should be *manageable,* given proper management attention.

CONCLUSIONS OF THIS ASSESSMENT

The analysis presented in this chapter leads to the following conclusions:

1. *If significant cost saving is the objective, transferring functions out of DNA is not the answer.* As long as the receiving organizations perform the functions at about the same levels of activity at which they are currently being performed by DNA, the savings will be modest. Substantial cost savings require substantial reductions in activity levels. If this is done, the cost savings will be about the same, whether the functions remain at DNA or are transferred elsewhere.

2. *If a decision is made to disestablish DNA,* there is no one organization to which all of DNA's current functions can be transferred without substantial risk. The preferred recipient organizations for its current functions are:

   - For the nuclear weapons stockpile support functions: a tri-service office reporting to OSD (first choice) or to STRATCOM (second choice).

   - For the NWE research and operational support functions: one or more of the DoE weapon laboratories.

   - For the nuclear threat reduction and arms-control functions: OSIA (first choice) or a split of functions between OSIA (nuclear threat reduction activities) and one of the DoE weapon laboratories (arms-control verification technology R&D).

   - For the conventional defense technologies functions: one or more of the service R&D organizations or ARPA.

In each of these cases, the risk (if any) to the maintenance of an essential function currently performed by DNA, if that function is transferred to the indicated organization, is at worst *manageable,* by which we mean that it can be effectively mitigated in a straightforward fashion that we have been able to identify, and in some cases only minimal. However, transferring these functions out of DNA will aggravate the fragmentation of the nuclear infrastructure, give nuclear weapon responsibilities to organizations with higher priorities, and will not by itself result in substantial cost savings.

3. *If the decision is made to continue DNA's existence,* the option of changing DNA into a Defense Special Weapons Agency has merit, provided management attention is squarely devoted to maintaining essential nuclear competencies within that agency.

A decision between the disestablishment of DNA (congressional Options 1, 3, and 4) and the continued existence of DNA (congressional Options 2 and 5) should not be made now, but hinges on larger considerations, related to the evolution of the overall U.S. nuclear infrastructure. We turn to this question in the next chapter.

Chapter Five

THE NUCLEAR INFRASTRUCTURE: ADDRESSING THE LARGER ISSUES

INTRODUCTION

The DNA functional issues RAND was tasked with addressing in this study, and which have been covered in Chapter 4, are important, but they reflect only one fairly small aspect of the larger, more complex problems associated with the direction, management, and core competencies of the nuclear infrastructure *as a whole* (i.e., the problems of downsizing and refocusing the entire U.S. nuclear infrastructure in the new post-Cold War environment).

The broader issues presented in this chapter emerged during the numerous interactions the study team had with diverse members of the extended nuclear community. These issues are obviously very complex, with technical, organizational, and political dimensions that will require a great deal of study in the coming months. The study did not analyze these issues in the same depth it studied DNA itself, and was not meant to do so. The RAND team offers the following observations and plausible hypotheses as *inputs* to this process rather than as *solutions.*

We begin this chapter with a profile of the U.S. nuclear infrastructure, as it evolved historically and as it exists today. Next we summarize how DNA fits into this infrastructure, and how its future is, or should be, linked to the evolution of the overall infrastructure. We then suggest a number of principles to follow in shaping the evolution of that infrastructure; these principles are guided by the enduring requirements identified in Chapter Two and by the stresses we see challenging the maintenance of those requirements in the post-Cold War world. Applying these principles to (1) that portion of the infrastructure within the DoD, and (2) the entire infrastructure, we develop two additional, broader options, beyond those on the congressional list. These involve, first, the consolidation of all nuclear infrastructure activities currently within DoD and, second, the consolidation within DoD of the entire U.S. nuclear infrastructure. We discuss the pros and cons of both of these options, and enumerate the many issues raised in particular by the second one, the consolidation of the entire U.S. nuclear infrastructure. We conclude the chapter by spelling out the implications for DNA of these broader options.

THE NUCLEAR INFRASTRUCTURE

Historical Background

The provisions for the management of nuclear weapons activities established after World War II were unprecedented and unique. The underlying authority for nuclear weapons was established (and is still found) in the Atomic Energy Act. Nuclear energy in general and the production of nuclear weapons in particular were the objects of intense congressional interest. The insistence on civilian control of nuclear energy resulted in the establishment of the Atomic Energy Commission (AEC). For a period of time, all nuclear weapons were controlled by AEC civilians. The total program was overseen by the Congress through the Joint Committee on Atomic Energy. Coordination with the military was effected through the Military Liaison Committee, supplemented by the Armed Forces Special Weapons Project (AFSWP), a predecessor to DNA. During the early era, all the military services wanted a nuclear capability, and the Congress, the AEC, and the White House supported their requirements.

Shortly after World War II a major new effort, led by DoD, was fielded in the area of nuclear weapon effects. The effort investigated the effects of nuclear radiation, thermal pulse, and blast on people, structures, military hardware, and communications systems. This effort created an important and complementary role for the DoD in the nuclear program—that of a developer and repository of knowledge about nuclear weapons effects and mitigation techniques. When this area of research was well established with extensive data accumulated, emphasis shifted to testing to determine the effects on military systems. The successor agencies to AFSWP, the Defense Atomic Support Agency (DASA) and later DNA, were the lead agencies for these vulnerability and hardening activities and played a major role through their contractors in carrying out these investigations.

In the late 1950s and early 1960s, as the nuclear stockpile became ever larger, the services were given ownership of the logistics and support organizations for their own nuclear weapons. Along with many different types of weapons, nuclear forces were deployed over a wide region of the globe and a short response capability was critical to our nation's defense. With large numbers of weapons of many different types deployed all over the world, safety and security were much more difficult to ensure; there was simply more to keep up with and more organizations were tasked, worldwide, to do it. In this era, DNA's predecessors, first AFSWP and later DASA, could no longer meet those demands, so the services developed their own organizations out of necessity.

As time went on, the need for centralized management and oversight of the nuclear infrastructure was perceived to be less important. The AEC and the Joint Committee were abolished. The nonregulatory activities of the AEC, including the nuclear weapons establishment, were assigned to the Energy Research and Development Administration (ERDA) and later to the DoE.

During the nearly 50 years since the detonation of the first U.S. nuclear weapon at the Trinity site, the U.S. nuclear stockpile grew from an initially small number of weapons of a few different types to a very large number of weapons of many different types. With the end of the Cold War, the number of weapons in the stockpile is now returning to the levels of the mid to late 1950s, and the number of different types of weapons is declining as well. Figure 5.1 (DNA, Field Command, RAND, 1994) illustrates this evolution.

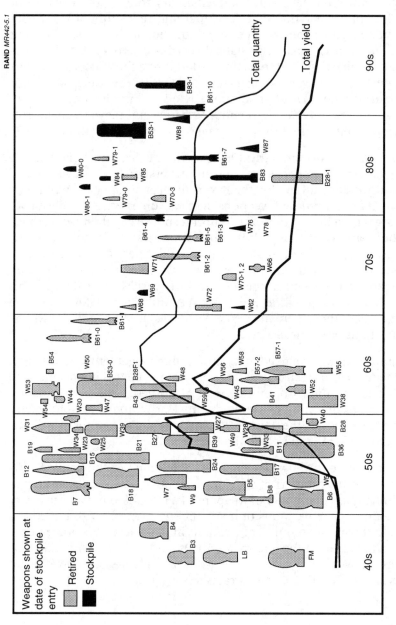

Figure 5.1—U.S. Stockpile Evolution

The Infrastructure Today: The DoE and DoD in Transition

To understand the organizational challenges facing the U.S. nuclear infrastructure, as it downsizes and refocuses on the demands of the new post-Cold War environment, we need a clear picture of the infrastructure as it exists today. The next several pages provide such a picture, in capsule form. We begin with an overview of the DoE and DoD nuclear missions, then briefly describe infrastructure activities during the various stages in the life cycle of a nuclear weapon. The stockpile phase is usually the longest part of this life cycle; we provide additional details on activities during that phase. Finally, we discuss current procedures for managing and coordinating infrastructure activities, and describe where DNA fits into all of this.

DoE and DoD Nuclear Missions. Today, the DoE and DoD are jointly responsible for the nation's nuclear deterrent. The DoD is responsible for nuclear requirements development, weapon system acquisition, deployment, operational planning, and employment and maintenance of nuclear weapons in the active and reserve stockpiles. DoD also conducts research and testing related to nuclear weapons effects to develop hardware and operational concepts that assure warfighting systems (nuclear and conventional forces) and supporting systems (communications and intelligence) are effective (survivable, reliable, operable) in nuclear threat environments.

There are, in effect, two principal interrelated DoD nuclear mission areas with shifting relative priorities and significance. The first is the maintenance of viable and credible *offensive* nuclear forces and support systems that underpin our deterrent posture. These nuclear forces are being reduced consistent with START agreements and no new weapons are currently under development, although it is clear that current service life extension programs will at some point run their course, necessitating the development and fielding of new systems to sustain the country's offensive nuclear deterrent essentially "forever." The second mission area is *survivability*, including hardening and other passive and active means for assuring the survivability and operability of our systems against nuclear threats. The character of these activities is changing as well, with a natural deemphasis on Cold-War threat issues (e.g., survivability and penetrativeness of nuclear offensive forces) and increasing emphasis on emerging threat issues associated with nuclear proliferation (e.g., the operability of conventional forces in a possible nuclear MRC). DNA has roles in each mission area, but their focus has been on the second, which is changing at such a rapid pace that it is challenging the nation's ability to respond with appropriate policies, programs, and guidance.

The DoE mission is to support the DoD by developing nuclear-weapons-related technologies, designing and testing nuclear warheads, producing these warheads (and their critical nuclear materials), helping the DoD in stockpile surveillance to assure the safety, security, and reliability of the warheads, and, finally, retiring the warheads and destroying or recycling their components. The physics and engineering design of the nuclear devices is done at Los Alamos National Laboratory and Lawrence Livermore National Laboratory. Sandia National Laboratory develops and fabricates the nonnuclear components of the weapon, including high-explosive triggers and related electronics. The current weapons production complex also includes plants at Pantex, Savannah River, and Kansas City.[1]

Infrastructure Activities During the Nuclear Weapon Life Cycle. The seven distinct stages in the nuclear weapon life cycle are:[2]

- *Concept formulation,* the development of warhead requirements for a new nuclear weapon system and preliminary concepts for achieving them in an actual nuclear warhead.

- *Feasibility,* involving tradeoff studies and analysis relating to the weaponization of the nuclear device and its effect on military operations, for a current or planned delivery system.[3]

---

[1]At the height of the Cold War, the weapons production complex also included facilities at Hanford (Washington), Rocky Flats (Colorado), Mound (Ohio), Pinellas (Florida), and in Idaho. These have all been (or are in the process of being) deactivated and refocused toward other activities.

[2]This discussion and the accompanying figure are based on material provided by the DNA Field Command (DNA, Field Command, 1994).

[3]"Feasibility" is somewhat of a misnomer for this stage in the life cycle, since it does not include any experiments required to demonstrate the physical or engineering feasibility of the nuclear device in question. In many cases, the conceptual design is based on previous nuclear devices whose feasibility is well established. Where it is not, the physical or engineering feasibility is normally established during the subsequent, development stage.

- *Development*, the actual physics design and preliminary engineering design of the nuclear weapon and (in former times) any experimental tests required to establish physical or engineering feasibility.

- *Engineering*, the detailed engineering design of the weapon.

- *Production*, the fabrication of the nuclear and nonnuclear components of the weapon, and their assembly and integration, in the quantities required.

- *Stockpile*, usually the longest part of the life cycle, during which the weapon resides in either the active or reserve stockpile.

- *Retirement*, when the weapon is removed from the stockpile and dismantled.[4]

Table 5.1 identifies the DoE and DoD agencies that participate in nuclear weapons infrastructure activities at each stage in the nuclear weapon life cycle. Areas of primary participation in a given stage of a nuclear weapon life cycle are shown by the symbol •; areas of secondary activities are denoted by the symbol o.

Perhaps the most critical elements of the nuclear infrastructure, at each of these life cycle stages, are not the structural and organizational elements discussed above, but the key core capabilities of its human resources, along with unique research, engineering, manufacturing, and production facilities tailored to nuclear weapon needs. Table 5.2 indicates, qualitatively, the areas where critical expertise has been developed to perform functions required at various stages of the nuclear weapon life cycle. Also shown are rough estimates of the number of people involved in different stages of the cycle at the height of the Cold War.

---

[4]This is becoming an increasingly important stage of the life cycle. The safe dismantlement of a nuclear weapon requires unique expertise in weapon design and engineering. Nuclear weapons necessarily are made of highly toxic materials that are rarely used elsewhere. The experience base on the long-term behavior of many of these materials resides only with the scientists and engineers within the weapon production complex, particularly at the weapon laboratories. Some weapons currently being dismantled are older than the engineers and technicians doing the dismantlement. In some cases, the original weapon designers are already retired. This can complicate the dismantlement process.

Table 5.1

Nuclear Weapons Life Cycle Elements: Primary Organizational Infrastructure

Participants	Life Cycle Stage						Stockpile	
	Concept	Feasibility	Development	Engineering	Production	Stewardship	Operational Use and Effect Testing	Retirement
DoE	●	●	●	●	●	●	○	●
HQ/field Site	●	●	●	●		●		●
Laboratories								
LLNL	●		●			○	○	○
LANL	●	●	●			○	○	○
SNL			●	●		○	○	○
Facilities								
Pantex					●			●
Savannah River					●			●
Kansas City					●			
DoD	●	○	○	○	○	○	●	○
OSD	●	○	○			○	●	○
Services	●	●				●	●	
Joint Commands			●				○	
DNA	○	○	○	○	○	●	●	○
NWC	○	○	○	○	○	○	○	○

Table 5.2

The Cold War Nuclear Infrastructure

Life-Cycle Stage	Participants	Example Areas Needing Critical Expertise	Cold War Infrastructure Size (Approximate)
Concept	DoE LLNL LANL Services DNA	Executive functions Requirements analysis Designers Device testers	1,000s
Feasibility	Services Joint Commands DNA	Executive functions Operations analysis	1,000s
Development, engineering, and production	DoE DoE weapon lab DoE facilities	Executive functions Material production Fabrication Production/manufacture Environment restoration QA/audits/inspections	100,000s
Stockpile stewardship	DoE Services DNA	Executive functions Incident/energy resource Safety and security Testing Maintenance/logistics	10,000s
Stockpile operations and effects issues	DNA Services Joint Commands	Executive functions Nuclear environments Effects of nuclear explosions on operational forces and systems Nuclear effects, simulation, and testing	10,000s
Retirement	DoE DoE facilities	Executive functions Material recovery Engineering experience Safety	1,000s

Infrastructure Stockpile Activities. As shown in Table 5.3, stockpile weapons are either: (1) integrated into operational units (CINC control and JCS reserves) and stored on alert delivery platforms (e.g., ICBMs and SLBMs) or in storage sites ("bunkers") near their delivery systems (e.g., bombs and air-launched cruise missiles (ALCMs) for

Table 5.3

U.S. Nuclear Stockpile

Operational Weapons	Reserve Weapons
CINCs—Operational	Active inventory
JCS—Reserves	Inactive inventory
Retired Weapons	**QA/RT Weapons**
DoD custody storage	DoE custody storage
DoE/DEMIL (Pits @ Pantex)	

Relative Numbers—Operational:Reserve:Retired = 1:2:5 (approximately)

the bombers); (2) held in reserve in service depots as active spares (e.g., suitable to replace operational weapons as required) or inactive spares (e.g., possibly cannibalized); (3) retired weapons scheduled for destruction (most stored in service depots, some by DoE at facilities such as Pantex); and (4) a few weapons within DoE's control that are periodically torn down and inspected, rebuilt, and replaced in DoD's stockpile.

Reserve weapons, both active and inactive, are stored by the services and are roughly twice the number of operational weapons. Retired weapons, currently the single largest category at perhaps five timesthe number of weapons in the operational inventory, are also stored by the services awaiting dismantlement.

Table 5.4 lists the DoD programs and responsibilities for maintaining the nuclear weapons stockpile. These have been organized into three functional areas: surety, logistics, and life cycle engineering support.

The DoD surety program is implemented by the services and DNA. In addition to DoD surety directives, each service implements its own unique directives. Each service performs inspections at its own nuclear weapons facilities to ensure adequate personnel reliability, and to insure that weapons systems and nuclear materials security and safety standards are being met. DNA performs inspections of its own, and in conjunction with the services, to establish an independent check for compliance with the service surety programs and DoD

Table 5.4

DoD Nuclear Stockpile Maintenance Responsibilities

DoD Nuclear Stockpile Surety Program
  Personnel reliability
  Weapons and material security
  Weapons safety
  Inspections
  Training
  Response
  Environmental restoration

DoD Nuclear Stockpile Logistics Program
  Weapons maintenance
  Weapons transportation
  Weapon storage
  Logistics management

DoD Nuclear Life Cycle Engineering Support
  Nuclear weapons survivability
  Nuclear weapons effects

imposed standards.[5] DNA has taken over from the individual services the responsibility for nuclear weapons training. This move, recently undertaken, was in response to the necessary downsizing of the nuclear stockpile and supporting personnel, and the recognition of the need for a single training program to serve all the services and CINCs. Training includes courses dealing with nuclear accidents, emergencies, explosive ordnance disposal, and general orientation on nuclear weapons and effects. Finally, each service with nuclear forces (namely, the Air Force and Navy) maintains an emergency action capability at differing locations to respond to accidents or incidents occurring at their nuclear sites. DNA also maintains an emergency action capability, with a command and communications center.

The DoD stockpile logistics program includes weapon maintenance activities, supporting inventory and supply systems, weapon storage facilities, and logistics management. Before 1991, the number of facilities, including submarines and ships, that possessed nuclear

---

[5]Each service's surety standards differ, often because of the differing environments in which they operate.

weapons exceeded 400. Currently that number is about 45, including 15 submarines, 6 major maintenance and storage depots, and 24 (13 overseas and 11 in CONUS) operational sites (e.g., intercontinental ballistic missile (ICBM) and bomber wings). Each operational site has its own nuclear weapons maintenance facility. The size and nature of the individual nuclear operational sites vary, but each requires many tens of guards, tens of maintenance personnel, and many tens of operational and management personnel. The services run six major storage depots (four by the Navy and two by the Air Force), estimated to cost between $20 million and $25 million each to operate annually (DNA, Field Command, RAND, 1994). Except for the requirement for the Air Force to transport nuclear weapons at the direction of the JCS, all off-base transport of nuclear weapons within the continental United States (CONUS) is performed by the DoE with a fleet of safe, secure trailers.[6] Thus, all nuclear weapons are delivered to the services for operational use and returned to DoE for dismantlement or repair by the DoE.

Finally, another area of stockpile support activities is in nuclear logistics management. There is a network of six service-run sites where material and parts for nuclear weapons systems are stored. This network originally served a much larger stockpile and could likely be downsized and consolidated. Of similar importance is a nuclear weapon readiness requirement that leads to the logistics "requirement" for storing 400 percent of estimated needs for limited life components and other unique spare parts. Whether the degree of readiness implied by this requirement will continue will in part depend on the outcome of the Nuclear Posture Review (NPR), currently being conducted by the DoD.[7]

Coordination and Management of Infrastructure Activities. Because responsibilities are spread between DoD and DoE, coordination of these activities is critical. Coordination among the participating organizations is done through the Nuclear Weapons Council

---

[6]The DoE transportation system was developed to prevent the theft of nuclear weapons during transit by bands of terrorists or other militant groups.

[7]This review will establish requirements and identify doctrine for the maintenance and employment of nuclear weapons. The NPR may also establish requirements for hardening and survivability of U.S. forces and systems that may be exposed to environments produced by nuclear weapons of other countries.

(NWC) and its associated committees. The NWC is an interdepartmental review and advisory body for coordinating nuclear weapon acquisition and related programming and budgetary matters. The NWC provides broad guidance regarding priorities and budget levels for research on nuclear weapons and nuclear weapons effects and oversees any other joint nuclear activities mutually agreed to by the DoD and DoE. The Under Secretary of Defense for Acquisition and Technology chairs the NWC with the Vice Chairman of the Joint Chiefs of Staff and the Assistant Secretary of Energy for Defense Programs as members. The NWC was formed following the Blue Ribbon Panel review of the U.S. nuclear weapons program in 1985 that concluded that the NWC's predecessor, the Military Liaison Committee with DoD and DoE participation, was ineffective, especially as its activities related to safety and security matters. (President's Blue Ribbon Task Group, Report, 1985).

Within the DoE, the Assistant Secretary of Energy for Defense Programs has the responsibility and authority for overseeing all phases of DoE participation in nuclear weapon life cycle activities. Unlike the DoE, the DoD does not have a single point of oversight, authority, or responsibility for its portion of the nuclear infrastructure, other than the Secretary of Defense. DoD responsibilities are now spread across several defense agencies, the services, and nuclear CINCs (e.g., USSTRATCOM). For example, within OSD the Assistant to the Secretary of Defense for Atomic Energy has the primary role for staff coordination on nuclear matters (as well as other WMD issues including some threat reduction and counterproliferation efforts). DNA "reports to OSD" through the ATSD(AE). However, budgetary authority and planning responsibility for DNA and other DoD agencies that are involved in nuclear issues, such as ARPA, OSIA, and the Armed Forces Radiological Research Institute (AFRRI), are held by the Director, Defense Research and Engineering (DDR&E), and by the services for their nuclear activities.

Where DNA Fits in Today's Infrastructure. As illustrated in Table 5.1, DNA plays a role in all stages of the nuclear weapon life cycle. DNA has been the DoD center of excellence for understanding and documenting the effects of nuclear weapons on systems—i.e., for vulnerability assessments and hardening support to System Program Offices (SPOs) and targeting support to the nuclear CINCs. DNA has developed, in conjunction with the DoE and service labs and service

SPOs, hardening technologies and design approaches that mitigate the effects of nuclear weapons. DNA has also developed and operated test facilities to validate the nuclear hardness of friendly systems and to assess the vulnerabilities of foreign systems.

In general, DNA is the principal operating interface between the DoD and the DoE in all field (nonlaboratory or production facility) activities involving nuclear weapons. DNA assists in developing nuclear weapons requirements and in assessing concept feasibility. DNA also assists in the translation of CINC operational needs into weapon design goals through an interactive process between the DoD and DoE. In the development, engineering, and production phases, DNA has some oversight responsibility for DoE's quality assurance program, which involves interactions with the production facilities and occasionally with the material suppliers.

DNA has an independent quality assurance role in the service's stockpile maintenance activities for all weapons in service custody, conducting independent audits of the individual service safety and security inspection programs. DNA represents the DoD in matters dealing with the logistics support of nuclear weapons by maintaining a database for the JCS with the status of all nuclear weapons in DoD custody and by managing the DoD-wide system to replace limited life components (LLCs) and other unique spare parts.

The functions that DNA now performs are highly cross-linked with those of the DoE and the services. In the areas of estimating and testing the effects of nuclear weapons on all operational forces and systems, DNA's involvement with the services, CINCs, and (to a lesser extent) the DoE is wide ranging. This high degree of integration between the functions now performed by DNA and those performed by other parts of the nuclear infrastructure underscores the need to set this assessment of DNA functions into a broader context related to the evolution of the nuclear infrastructures as a whole.

This completes our capsule tour of the U.S. nuclear infrastructure. As we have seen, it started in a very centralized form immediately after World War II and became increasingly decentralized as it grew in size. As a consequence of this decentralization, a complex arrangement evolved over time, with a number of different organizations having interdependent responsibilities. With this picture as a point of departure, we now turn to the organizational challenges confronting the infrastructure as it adapts to post-Cold War priorities.

ADAPTING TO POST-COLD WAR PRIORITIES

A graphic measure of the post-Cold War realities is seen in the budget and manpower cuts that have been made that affect the nuclear infrastructure. Figures 5.2 and 5.3 provide two illustrations of these trends, which are likely to continue. Shown in Figure 5.2 is the steep decline in recent years in the annual DoE budget for Defense Programs, which includes nuclear weapon R&D, production, stockpile stewardship, and dismantlement. There are currently *no* new nuclear weapons being designed or produced by DoE. Activities in support of stockpile maintenance—to maintain confidence in the safety, reliability, and performance of U.S. weapons in the absence of

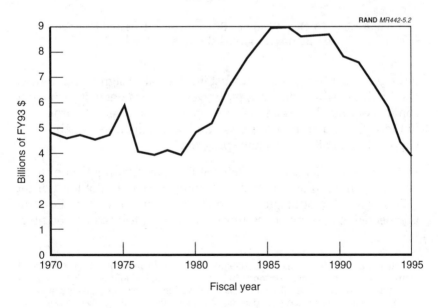

Figure 5.2—DoE Budget for Defense Programs

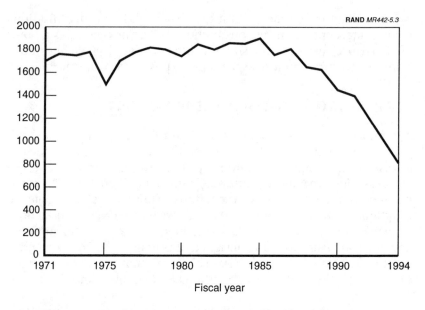

**Figure 5.3—Nuclear Weapons Program Scientists and Engineers,
Los Alamos National Laboratory (1971–1994)**

underground testing[8]—and to sustain nuclear design competencies within the labs are under way, however. These activities include some continued design efforts, new experimental programs, numerical simulation programs, and a production capability to support a stockpile surveillance program.[9]

As Figure 5.2 indicates, DoE Defense Program funding began to decline beginning in about 1989, returning by 1994 to the level of the 1970s, before the Strategic Modernization Program began. Funding is expected to continue declining over the next few years. The num-

---

[8]Nuclear weapons in the existing stockpile will have to remain safe and viable for many years, in some cases beyond their original design lifetimes. New technical problems or problems with the materials employed in nuclear weapons are expected to emerge that will require new and acceptable methods for testing weapon safety and reliability.

[9]As part of DoE's reconfiguration program, a small nuclear weapon production capability is scheduled to be installed at Los Alamos and a small components production capability at the Sandia National Laboratory to support the stockpile stewardship program.

ber of scientists and engineers engaged in nuclear weapons R&D is following suit.   For example, Figure 5.3 (Sigmund, *Background Information*) shows the number of scientists and engineers at Los Alamos National Laboratory engaged in the nuclear weapons program.[10] The number of people directly involved in nuclear R&D has been declining since about 1986.

Budgets and manpower profiles within the DoD are distributed among several organizations and are less easily aggregated than in the DoE.   However, one indicator of the response to post-Cold War drawdown is the DoD spending on strategic nuclear forces.   Figure 5.4 illustrates these expenditures.   As the stockpile is reduced substantially, DoD costs are also being reduced.   As mentioned above, today the services operate about 45 sites for stockpile weapons storage or maintenance; about 400 such sites were in use in early 1991. For many years, the services supported their individual nuclear

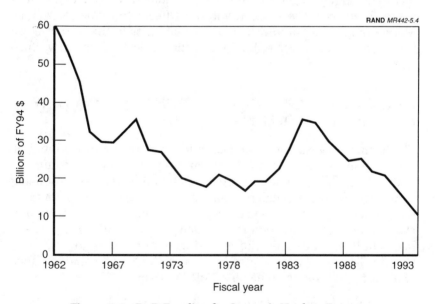

Figure 5.4—DoD Funding for Strategic Nuclear Forces

[10]Lawrence Livermore Laboratory's Institutional Plan (Lawrence Livermore, 1993) describes a similar trend at Lawrence Livermore National Laboratory, where the total number of people involved in the "core" nuclear weapons program will continue to decline.

weapons programs with R&D within their respective SPOs, laboratories, or arsenals. Today, these service nuclear-weapons-related activities have been severely downsized. The DoD nuclear community is shrinking, with DNA the sole remaining DoD agency with a focus on nuclear RDT&E activities.

As appropriate with the shrinking of the nuclear stockpile, the number of nuclear-rated officers and enlisted personnel in the services is decreasing, as is the number of civilian laboratory and industrial base personnel with critical expertise. Figure 5.5 indicates recent manpower trends in the Army, Navy, Marines, and Air Force for nuclear-rated personnel.

The picture is clear: The funding available to support the U.S. nuclear infrastructure has decreased substantially in recent years, and will most likely continue to decrease; the number of civilian and military personnel engaged in infrastructure activities has declined substantially in recent years, and will most likely continue to decline. Maintaining essential infrastructure capabilities in the face of these budget and manpower reductions is the subject to which we turn next.

REDUCING THE INFRASTRUCTURE IN A RATIONAL WAY: PRINCIPLES TO FOLLOW

The further downsizing of the U.S. nuclear infrastructure should be based upon careful tradeoffs between future needs and sustainable budget levels, to maintain essential capabilities and avoid producing a number of disconnected, possibly redundant pieces, many of them with subcritical levels of expertise. In shaping the downsizing and evolution of the infrastructure, it is helpful if a set of principles can be identified, guided by the enduring requirements identified in Chapter Two and by the stresses we see challenging the maintenance of those requirements in the post-Cold War world.

Chapter Two identified three important, continuing, nuclear-related requirements the United States must face:

• A diligent stewardship of the nuclear stockpile, to include the safety, security, and reliability of U.S. nuclear weapons, and possible modernization needs (many of them safety-related).

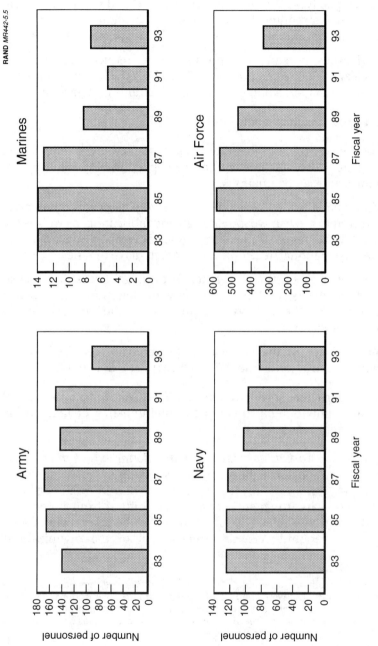

Figure 5.5—Manpower Trends in Number of Army, Navy, Marine, and Air Force Nuclear-Rated Officers

- The capacity to understand and deal with the use of nuclear weapons (and other WMD), to include accidental or intentional detonations and their effects on U.S. interests, allies, and forces.

- A vigorous pursuit of opportunities to reduce the threat of nuclear weapons and other WMD use by reducing the number of weapons that exist and the number of nations that have them.

Fulfilling these requirements implies:

- Continued availability of the specially trained, experienced, and diligent personnel needed for stockpile inspections, maintenance, and safety functions.

- Maintenance of the core physics and engineering competencies involved in nuclear weapon design and fabrication, as well as the capability to reconstitute a larger nuclear weapon inventory if future international developments were to make that necessary.

- Maintenance of the necessary knowledge bases in nuclear weapon effects, and the capability to determine the nuclear hardness of friendly systems and assess the vulnerability of foreign systems.

- Maintenance of the verification technologies required for effective nuclear (and other WMD) threat reduction activities.

As the infrastructure shrinks in the post-Cold War world, a number of interrelated phenomena threaten the maintenance of these infrastructure capabilities:[11]

- *The basic individual and organizational urge to survive.* As a result of straightforward organizational and individual survival responses to the reductions currently under way, nuclear-weapon-related activities are shifting from being very high priority to relatively low priority—to be taken care of after almost everything else is done—at all of the organizations

---

[11]These phenomena are not unique to the post-Cold War era; they have existed as long as human activities were conducted in large, complex organizations. What the drawdowns associated with the end of the Cold War have done is accentuate the stresses that these phenomena place on the nuclear infrastructure (as well as on other parts of the U.S. military-industrial complex).

involved in the nuclear infrastructure, almost without exception. Almost all the individuals and organizations believe that their future—success, prosperity, well-being, etc.—lies in the non-nuclear realm and are "diversifying" into that realm as rapidly as they can.

- *The difficulty of attracting new blood and fresh faces to a non-growth, apparently low-priority area.* Maintaining essential infrastructure capabilities for an extended period of time (measured in decades, not years) requires more than the mere retention of some fraction of the skilled personnel currently in the nuclear infrastructure; it requires the recruitment of new people, with the proper talents and skills. In each generation, the best and the brightest people tend to go into those areas to which society appears to assign a high priority and which have prospects of growth, not into an area to which even the organizations involved appear to assign low priority.

- *The difficulty of maintaining a critical mass in specialized functions, in the individual pieces of a dispersed organization undergoing downsizing.* As the many individual pieces of the nuclear infrastructure shrink, many of them could fall below critical mass, in terms of the quantity and quality of necessary expertise and resources. As the management literature demonstrates, during downsizing specialized capabilities can readily be lost altogether.

- *The difficulty of detecting and correcting the inevitable capability gaps that develop in a fragmented organizational structure.* As a result of the "survival responses" described above and the difficulties of maintaining critical mass in a dispersed structure, some capability gaps may well occur. Because of the fragmentation of the current infrastructure, no individual or set of individuals has the visibility across the breadth and into the depths of the infrastructure necessary to detect and correct these gaps in a timely fashion.

- *The difficulty of properly allocating and focusing resources in a fragmented organizational structure.* Also because of the fragmentation of the current nuclear infrastructure, no one individual has the purview or authority across the breadth and into the depths of the infrastructure necessary to efficiently allocate limited resources, ensuring their focus on the most important tasks. As the infrastructure shrinks, this will almost inevitably lead to a series of suboptimizations, with the inherent inefficiencies that these will entail.

Taken together, these phenomena raise the prospect of a future nuclear infrastructure with a number of inefficiencies and, more important, with hidden capability gaps in the maintenance of essential functions.[12]

In coping with these phenomena, we suggest the following *two principles,* as guides to the continued downsizing and evolution of the nuclear infrastructure in a rational manner:

- **The value of priority.** *Nuclear-related activities are no longer the highest priority for the nation, but given the special character of nuclear weapons and the uncertainties of the future, they should still be high priority for the organizations to which they are entrusted.* People and organizations do well the things to which they assign a high priority. They can be depended upon to perform high-priority tasks adequately. This is basic human and organizational nature. The most straightforward way to combat the stresses listed above is by designing an organizational structure for the downsized nuclear infrastructure in which the key functions supporting the essential requirements reside in organizations in which they receive high priority.

- **The efficacy of consolidation.** *The most straightforward way to ensure that the key nuclear-related functions reside in organizations in which they receive high priority is by consolidation:* a bringing together of functions currently dispersed across a wide variety of organizations, many of which treat them as lower-priority tasks, not central to the future vitality of the organization. Besides ensuring the priority attention necessary for confident maintenance of these functions, consolidation also makes the maintenance of critical masses in specialized functions, the detection and correction of any capability gaps that do occur, and the efficient allocation of resources much easier and more

---

[12]This is a prospect, not a certainty, but it is a prospect that a prudent nation would do well to avoid.

straightforward. Consolidation is important both for managing more effectively under reduced resources and in going through the process of change and redirection.[13]

We will now apply these two principles, first to that portion of the nuclear infrastructure within the DoD, and later to the entire U.S. nuclear weapon infrastructure.

CONSOLIDATING DoD FUNCTIONS

As our discussion earlier in this chapter suggests, the portion of the nuclear weapon infrastructure within the DoD is a prime candidate for future stress-related "illness" of the type considered here. The DoD portion of the infrastructure is scattered across a number of service organizations and defense agencies, almost every one of which is rapidly diversifying into nonnuclear areas and, inevitably, downgrading the priority it attaches to its nuclear activities and re-sponsibilities. As these individual organizations and agencies downsize their nuclear functions and diversify into other areas, in-dividual pockets of essential, specialized capability may well fall be-low critical mass and be lost. Because of fragmented lines of re-sponsibility and authority, no individual in DoD, other than the Secretary and Deputy Secretary of Defense, has the management purview necessary to allocate limited resources efficiently across this infrastructure, ensuring their focus on the most important tasks.

Given these "symptoms," the prognosis is for inefficiencies in the al-location of resources (of both money and people) and unanticipated capability gaps in the maintenance of at least some essential func-tions. The prescription—applying our twin principles of *the value of priority* and *the efficacy of consolidation*—is straightforward: Con-solidate the various DoD nuclear infrastructure activities, under one senior federal executive in OSD, and within one or at most a few agencies reporting to that executive. This executive should have overall management, planning, and budget authority for all of the

---

[13]The principle of consolidation does not mean that organizations should work *only* on nuclear-related matters, to the exclusion of everything else. They can well work on other things, so long as their nuclear activities are a large enough part of the whole to receive high-priority attention.

DoD nuclear infrastructure activities. The agencies involved should, as a result of the consolidation, have nuclear-related activities as such a significantly large portion of their total efforts that they will naturally assign them high priority.

This DoD consolidation could and should include the three DoD stockpile programs and responsibilities discussed above: the nuclear stockpile surety program, the nuclear stockpile logistics program, and the nuclear life cycle engineering support program. This consolidation could also encompass the executive responsibility for implementing the Cooperative Threat Reduction Program and other arms-control-related activities within OSD. Consolidated management for all nuclear support matters within DoD could include budgetary authority over DNA, OSIA, and AFRRI, oversight and review of service stockpile operations and related R&D activities, and a sufficient charter to effect essential consolidations of service facilities, the standardization of safety and security measures across the CINCs and services, and the ability to take possession of elements of the stockpile as agreed to by OSD and the services. One objective of this move would be to achieve cost savings from the consolidation of stockpile-related facilities and from effective planning, oversight, and resource allocation for all nuclear-related support activities throughout the DoD.

The services were originally given ownership of the logistics and support organizations for their nuclear weapons in the late 1950s and early 1960s, as the stockpile increased in size beyond the capabilities of AFSWP (and later DASA) to meet the demands for logistics and support. As the post-Cold War stockpile shrinks back to its size in the mid to late 1950s, the time has come to return to a more centralized support organization. We believe that this consolidation would provide a significant improvement in DoD nuclear infrastructure management, would reduce a number of the stresses threatening the preservation of essential functions, and could reduce costs significantly.

THE BROADER DEFENSE CONSOLIDATION OPTION

During the course of this study, which originally focused on DNA, and then expanded to consider the entire DoD portion of the nuclear weapons infrastructure, we had extensive interviews with organizations and individuals across the spectrum of U.S. nuclear weapon activities. As a result of these interviews, an even broader issue gradually emerged, the state of health of the entire U.S. nuclear infrastructure. There was a strong community concern voiced by the services, the national laboratories, and the private contractors about the loss of critical expertise within the nuclear infrastructure as a whole. It did not seem to these members of the nuclear community that the contraction of the nuclear infrastructure was coherent or coordinated or had adequate high-level visibility. These concerns cannot be dismissed as just being based on self interest; they must receive serious consideration.

A hypothesis that needs to be tested is that *there no longer are any important, enduring interests served by keeping responsibilities for the nuclear infrastructure split between the DoD and DoE.* A more consolidated infrastructure would permit a systematic set of tradeoffs to be done and actions to be taken providing at least five advantages in the management of a community that must be carefully scaled back from a considerably larger size. These include: (1) a sufficiently complete view of the infrastructure to head off potential problems before they occur, (2) the ability to seize potentially significant opportunities for cost saving and the ability to implement measures to realize such saving, (3) a greater chance of managing the risks of accidental loss of critical expertise posed by downsizing, (4) a more effective advocacy process in budget priority decisions, and (5) the ability to identify potential areas of synergism on an infrastructure-wide basis, combined with the authority to best use existing resources.

Logically, a consolidation of nuclear programs can be done either under the DoD or under the DoE. On balance, since some nuclear-related activities—those directly associated with delivery systems and operational employment—*must* remain closely coupled to the military services, the DoD option appears clearly preferable.

The principal elements of the U.S. nuclear infrastructure could be combined in a new DoD entity. The entity would encompass the key nuclear functions of DNA, the Defense Programs organization of the Department of Energy, and the Office of the Assistant to the Secretary of Defense for Atomic Energy (ATSD(AE)). Those portions of the DoE weapons laboratories responsible for nuclear weapons research and development also would be included in this new DoD organization.

At the headquarters level, the management functions of DNA, DoE Defense Programs, and the ATSD(AE) could be consolidated with likely savings in personnel. In the field, the functions of DNA Field Command and portions of the DoE Albuquerque Operations Office would also be candidates for consolidation. Since virtually all nuclear military activities would be centralized in DoD, the need for and role of liaison bodies between and within the departments could be reduced or eliminated.

Programmatically, the "core" programs at DoE in nuclear weapons research and development, as well as nuclear weapons stockpile management, would fall under this new DoD entity. All stockpile surveillance, training, transportation, maintenance, production, and conversion would be under the new DoD entity. Other DoE programs such as classification/declassification, safeguards, security, and nuclear weapons materials might be included as well.

Nuclear weapons responsibilities of the military services could be tied more tightly to a central organization. At present, limited-life components are exchanged by the services on their own weapons. This practice could continue, or it could be done by the new DoD entity for the services. Service training could be realigned, and DoD nuclear schools established. Combining nuclear and other WMD functions would be more compelling in the context of a larger (DoD-DoE) infrastructure than now, since the risk of a diversion of attention, resources, and talent away from nuclear weapons would be less.

ISSUES ASSOCIATED WITH THE BROADER CONSOLIDATION

Consolidation of the overall U.S. nuclear infrastructure is not a new idea, but it is an idea worth a new look. The shift and uncertainty in the world nuclear environment, the declining U.S. stockpile, the surplus of special nuclear materials, and declining budgets all suggest that consolidation could be given thorough consideration. However, such a decision would need to confront several large issues, and an effective consolidation plan would need to overcome significant obstacles. We here attempt a brief enumeration and discussion of these issues and obstacles.

New Bureaucracies Are Less Predictable Than the Old

It is reasonable to be leery of creating a new bureaucracy, within the DoD or elsewhere. Although there appear to be advantages in consolidating nuclear weapons functions, what is not desirable is the creation of a large entity inclined toward strong advocacy of better and more nuclear weapons. It would be essential for the designers and managers of a centralized entity to accept the declining scale and role of nuclear arsenals and to focus on the skillful management of this desirable decline.

Real Consolidation Is in the Field

The bulk of U.S. spending on nuclear weapons functions is in the field, at DNA Field Command and DNA contractors, and at DoE field offices, plants, and laboratories. Thus, to achieve savings and effectiveness, as well as improved management and oversight, the feasibility of bringing the field organizations under a common umbrella must be addressed.

Alternative Futures for the DoE Weapon Laboratories

As is pointed out in the Institutional Plans for the laboratories (Lawrence Livermore, 1993; Los Alamos, 1993; and Sandia, 1993), DoE has had difficulty supporting its weapon laboratories in recent years. Increasingly, DoE is turning to commercialization of its technology and work for other government agencies to help maintain the staffing levels at these laboratories. The laboratories themselves are receptive to new missions and to opportunities to apply their capabilities to a variety of important national problems. Even the DoE plants and field offices have followed this trend, diversifying and branching out into new areas.

Conceptually, several options would permit the assumption by DoD of the nuclear weapons activities of the DoE labs. One would be for DoD to assume management and budget responsibility for Lawrence Livermore, Los Alamos, and Sandia, and then to support DoE's non-weapon needs from the labs (in the sense that the labs now support DoD's weapons needs). Another would be to leave all three labs under DoE but give DoD direct management control and budget responsibility for distinct nuclear weapons programs performed at those labs (making DoE, in effect, a landlord). A third option would be to consolidate the nuclear weapons activities in one or two labs (e.g., Sandia and Lawrence Livermore or Los Alamos) and transfer them to DoD while concentrating DoE's nonweapons endeavors in the remaining DoE labs. We have no analytical basis within this study for indicating which option would be best, nor how difficult each would be to implement.

Headquarters Organizational Issues

If a new DoD organization were to consolidate most of the federal effort in the nuclear weapons area, the purposes of the Nuclear Weapons Council, the Nuclear Weapons Council Standing Committee, and the Nuclear Weapons Safety Committee would all need to be reviewed.

Several senior federal positions—notably the DoE Assistant Secretary for Defense Programs, the OSD Assistant to the Secretary of Defense for Atomic Energy, and the Director, DNA—would need to be combined, recast, or brought together under a more senior position. A number of functions would need to be considered for transfer to a new DoD organization, or combined with existing DoD functions. These functions would include: the production of special nuclear materials, including tritium, limited-life weapon components, research and engineering, transportation of nuclear weapons and materials, emergency response, and, some, if not all, of the nuclear security and classification programs. Other national-security-related functions that could be transferred to DoD could include DoE programs in intelligence and arms control.

The DoE Office of Nuclear Energy has certain defense-related nuclear responsibilities, other than nuclear weapons: the Office of Space and Defense Power Systems and the Office of Naval Reactors. There may or may not be a strong case to shift these functions to

DoD as well, as part of a move to consolidate nuclear weapons functions; we have not considered this issue. There should be some examination of whether it would be wise to leave only these defense activities in DoE.

Jurisdictional Questions

With a new DoD organization managing substantial programs at what are now DoE labs and plants, jurisdictional issues would arise at those sites, since DoE would be expected to have a continuing presence. Conflicts could arise over environmental cleanup issues, over liability and indemnification issues, or over divergent programmatic priorities and funding. For example, a former DoE weapons laboratory would presumably continue to be funded by DoE for some civilian R&D in, say, fusion, basic energy science, or biotechnology. If that same laboratory were managed and funded for its defense work by a new DoD organization, conflicts over those issues listed above could arise.

Environmental Restoration and Waste Management

Because of the very large scale of DoE environmental restoration and waste management activities, this is an especially difficult area. DoE has created the office of the Assistant Secretary for Environmental Restoration and Waste Management. This office consolidates nearly all of the DoE efforts in environmental cleanup and waste management, whatever the geographic or functional source of the waste.

Although these functions could continue to reside in DoE, it should be noted that they are funded virtually entirely out of defense function 050 funds. Also, under the assumption that DoE continued to manage environmental cleanup, there would arise the issue of who was responsible for *new* environmental problems created by a new DoD organization. It is not clear that bifurcating responsibility for nuclear waste cleanup—between old and new, or between that from weapons programs and that from other sources—would be prudent. Options concerning responsibility for this national priority would need thorough analysis.

Congressional Jurisdiction

Currently, DoE reports to a number of different congressional committees—a striking contrast to the days when the Joint Committee on Atomic Energy provided sole oversight of the AEC. DoE Defense Programs reports to the House and Senate Armed Services Committees and to the House and Senate Energy and Water Appropriations Subcommittees. Other congressional committees, e.g., the House Committee on Energy and Commerce, the House Committee on Space Science and Technology, and the Senate Committee on Energy and Natural Resources, have some jurisdiction over the DoE labs and the research programs they conduct. A new set of alignments would have to be clarified, taking into account these numerous interests. Finally, a reorganization of the federal government falls under the jurisdiction of the House Committee on Government Operations, and the Senate Committee on Government Affairs.

Civilian Control

Civilian control of nuclear weapons has been an issue from the beginning of nuclear weapons development. In 1974, when the Energy Research and Development Administration was established, and in 1977, when the Department of Energy was established, the importance of civilian control was reaffirmed by the Congress. Creating a new DoD entity that consolidates the U.S. nuclear weapons establishment will raise the issue of civilian control again.

The imperative of civilian control can be satisfied in many new ways. Today's civilian hierarchy in the Pentagon, in the laboratories, and in federal field offices all establish broad civilian control in a way not conceivable in the days of the Manhattan Engineering District. Further, budget pressures and new missions for the military services and the DoD have shifted priorities away from nuclear weapons. Within the military, if not the DoD as a whole, interest in nuclear weapons and in policy and resource discretion associated with them is different, and less than it was in the past. Most important, the consolidation in DoD of the nuclear weapons infrastructure under the management of OSD, as opposed to continued decentralization or a shift of functions to the services, should satisfy concerns about civilian control.

THE IMPLICATIONS OF THESE OPTIONS FOR DNA

The implications for DNA of these two consolidation options is profound. We have noted that DNA is a small part of a far-flung infrastructure and that dismantling or reorienting DNA is best considered in the context of a new design for that larger infrastructure.

One drawback of dismantling DNA is that it could impede the consolidation of DoD and DoE nuclear weapons functions inside DoD.[14] That said, it is not necessarily the case that DNA should be left unchanged in the context of such a consolidation, let alone that DNA would become the recipient of functions being centralized.

Although the question of addressing DNA's functions in a larger framework requires more analysis, a helpful conceptualization is to consider organizing a new entity to fulfill the three main requirements that are most important for the future: stockpile support, weapons effects and other RDT&E, and nuclear threat reduction and arms control. This would provide an organizing principle not only for the disposition of DNA nuclear functions, but also for those transferred to DoD from DoE. Such an approach is illustrated below in Figure 5.6.

RELEVANT EXECUTIVE AND LEGISLATIVE ACTIONS REGARDING THE NUCLEAR INFRASTRUCTURE

RAND is not the only or even the first party to suggest a realignment of the U.S. nuclear infrastructure. These dynamic but uncertain infrastructure issues have been the subject of much recent attention by the administration, the Congress, and both the DoD and DoE. At least a half dozen congressional bills are in formulation and studies are under way that would affect the future of DoE's nuclear programs, its weapons laboratories, or their reporting relationships. There have also been proposals in legislation for a Laboratory

---

[14]Shifting DoE nuclear weapons functions to DoD could be more difficult if the only significant DoD entity concerned mainly with nuclear weapons had been eliminated.

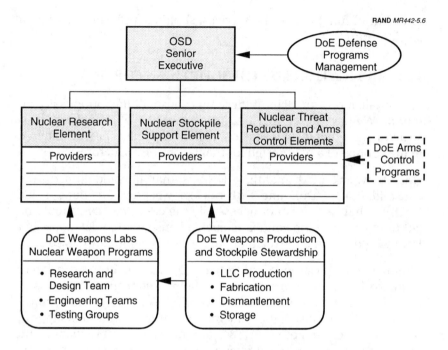

Figure 5.6—DoD Nuclear Consolidation Concept

Closure Commission, analogous to the Military Base Closure Commission, as well as congressional report language requesting studies and evaluations of the DoE weapons laboratories, the DoE Defense Programs organizations, and their collective futures. Another review is under way by the Nuclear Facilities Safety Board, commissioned by the Congress to review, over a five-year period, the safety and environmental cleanup performance of the DoE at its production and research facilities (Defense Nuclear Facilities Safety Board, 1993). The board has been critical of the DoE's performance in environmental cleanup at a number of locations and its recommendations could call for the closure of some DoE facilities and the transfer of some nuclear weapon activities to the DoD (Defense Nuclear Facilities Safety Board, 1993).

The initial findings of several other studies currently under way suggest that it would be safer and much more effective to consolidate our nation's nuclear weapons support functions under one agency.

The Fail-Safe and Risk Reduction Study chaired by Ambassador Jeanne Kirkpatrick (NSS-096-92), stated that in the current environment the United States must focus on simplifying and ensuring strong, consolidated control of our nuclear weapons stockpile. These study activities do not agree on how consolidation should be achieved or under what organizational authority. However, all endorse consolidation as a common management principle to mitigate the risk of an unsafe stockpile or one with weapons ill-matched to future deterrence needs.

Concurrently, the Department of Defense is leading a major Nuclear Posture Review (NPR) to be completed in late 1994. This review will stipulate the requirements and identify doctrine for the maintenance and employment of nuclear weapons. The NPR may also establish requirements for hardening and survivability of U.S. forces and systems that may be exposed to environments by nuclear weapons of other countries.

Finally, a special subpanel of the Secretary of Energy Advisory Board (SEAB) studied the DoE weapon laboratories from 1990 to 1992 and a new SEAB subpanel began a new study in March of this year (Sherwood, March 3, 1994) and is expected to complete its work by next year. Within the Executive Branch, OSTP is conducting a study of the federal laboratory structure. Both NASA and DoD are beginning studies of their laboratories as well. All three studies (NASA, DoD, and DoE) will serve as input to the OSTP review of the federal laboratory structure.

Although some of these activities are complete, others are ongoing and more are still pending. A clear path toward a new nuclear infrastructure for the nation has yet to be paved.

SUMMARY

The U.S. nuclear infrastructure is going through a period of profound change and increasing stress. These stresses could threaten the maintenance of infrastructure capabilities still essential for U.S. national security in the post-Cold War world. The most straightforward way to combat these stresses and ensure the long-term maintenance of these capabilities is through consolidation: consolidation first of all of the infrastructure functions and activities currently within the

DoD, and more broadly, consolidation of the entire U.S. nuclear infrastructure within one management entity, preferable within DoD.

The first consolidation, of infrastructure functions and activities currently within the DoD, is relatively straightforward and should receive serious and immediate attention. The broader consolidation, of the entire U.S. nuclear infrastructure, raises a number of large issues, and would need to overcome significant obstacles. These issues are obviously very complex, with technical, organizational, and political dimensions requiring a great deal of study. To repeat what was said above, in view of the limited nature of this study, the observations and hypotheses offered here regarding the broader consolidation should be considered as *inputs* to this process, rather than as *solutions.*

Given this caveat, but also given the significance of this subject to the future security of the nation, the findings of this study of DNA functions include a suggestion that the consolidation of all U.S. nuclear-weapons-related activities within the DoD be seriously considered as a primary organizational option for an enduring and robust U.S. nuclear infrastructure for the 21st century.

Chapter Six

FINDINGS AND RECOMMENDATIONS

Our analysis of DNA, the continuing future nuclear requirements, and the national nuclear infrastructure lead to a number of findings, from which we make three major recommendations for the study. These appear below.

KEY FINDINGS

• The range of uncertainty about the role of nuclear weapons in the future national security environment is great—from a benign world of small arsenals (100s) held by a few responsible states to a menacing world of widespread proliferation and large arsenals (1000s) possessed by unfriendly powers. Although U.S. interests are served by working vigorously for the former, the latter is conceivable.

• Although the threat to U.S. national survival is significantly less than it was during the Cold War, the likelihood that nuclear weapons will be brandished or used by a state hostile to the United States may be greater in the future than it was during the Cold War.

• The requirements imposed by this perspective of the future stem from the proposition that the United States should try to shape the nuclear future while maintaining the ability to cope with the most immediate environment and the potential to respond to the worst environment. Nuclear weapons present unique challenges and should be addressed explicitly in our defense planning.

- Under no circumstances should the United States find itself inferior to any other nuclear power or unable to function militarily in a nuclear environment. This suggests that the United States should have the capability to maintain qualitative and quantitative nuclear advantage and the ability to project conventional power in the face of nuclear threats.

- Thus, the U.S. nuclear infrastructure should be able to meet three basic requirements:

  — Maintain a safe, secure, and reliable stockpile (and improving it if necessary, as with safety-related changes);

  — Understand the effects of nuclear weapons used by or against U.S. forces and apply that understanding to support conventional operations in nuclear threat environments;

  — Have the means to pursue effective and ambitious arms-control and disarmament measures to dispose of excess nuclear weapons and material.

- The U.S. nuclear weapons infrastructure must be able to meet these needs. It is currently split between DoD and DoE and is further fragmented within both departments. Budgets and numbers of qualified staff are declining across the board, as they should in light of the reduced number of weapons, threat, modernization needs, and military role of U.S. nuclear weapons. However, the highly decentralized structure that was tolerable, perhaps preferable, during the Cold War presents risks for the future. In particular, there is now a greater danger of falling below critical mass and thus endangering our ability to meet future requirements. There is as well a danger that organizations for which nuclear weapons are no longer a main responsibility and a high priority will in fact neglect them.

- Within this fragmented infrastructure, DNA remains focused largely on nuclear weapon matters. Most of DNA's current functions and budget address the three critical future requirements. At the same time, DNA is increasingly active in a variety of nonnuclear weapon technology areas.

- Analysis of alternative ways of performing the functions of DNA revealed the following:

— Cost savings achieved by transferring all DNA's functions would be between $10 million and $20 million per year. Because DNA is primarily an executive agent, its overhead is a small fraction of a budget that supports functions that would have to be picked up by the receiving organizations.

— DNA has instituted procedures aimed at reducing manpower and overhead costs and is evolving into a "leaner DNA." These reductions are equivalent to an annual savings of at least $20 million once they are fully implemented.

— There is no single organization in DoD or DoE that could take on all of DNA's functions without creating substantial risks in some functional areas.

— However, if the DNA functions are distributed among the military services, ARPA, DoE, and OSIA in an optimum fashion, the risks are manageable in all areas, and in some cases only minimal.

— There are, however, several pitfalls in this case:

  • It could increase fragmentation.

  • It would place certain critical nuclear tasks within organizations that have higher priorities.

  • It might prejudice, if not impede, potentially important steps to reorganize the larger U.S. nuclear infrastructure for the new era.

• Indeed, serious consideration should be given to this larger national nuclear infrastructure issue. It is unclear whether there remains any important, enduring interest served by keeping responsibility for the nuclear infrastructure split between DoD and DoE. Merging these responsibilities in a carefully considered way could place the United States in a better, more confident position to meet key nuclear requirements at lower cost. Merger under DoE would be unwise, given DoD's operational nuclear responsibilities and its direct concern with nuclear effects and with threat reduction. But merger under DoD merits thorough examination, which must include feasibility, extent of the transfer, effects on remaining functions of DoE labs, and implementation plans.

- Consolidation of current DoD nuclear-weapons-related support activities (stockpile support, nuclear effects research, and threat reduction and arms control) would be a good first step in this direction, and useful in its own right.

- In the meantime, dismantling of DNA could make such a larger solution more difficult; shifting DoE nuclear weapon functions to DoD could be more complicated if the only significant DoD entity concerned mainly with nuclear weapons had been eliminated. Conversely, changes in the structure and scope of DNA for the long term are better made in the context of a blueprint for the national nuclear weapons infrastructure as a whole.

- Thus, for now, the nuclear weapons functions of DNA ought to remain in DNA (and their directions and funding assessed and set on their own merits). The marginal cost of leaving DNA intact for now is negligible.

- Disposition of the nonnuclear functions of DNA are, however, a different and more ambiguous matter. Although it is true that other WMD, counterproliferation acquisition planning, and conventional effects activities are potentially a way to help sustain resources and skills needed for nuclear weapons functions, there is a risk that these "growth" areas could divert management attention, resources, and superior technical talent. The case for combining biological and chemical weapons effects functions with DNA's nuclear effects functions is the easiest to make, but requires a clear indication that DoD management is aware of the pitfalls. Consolidating responsibility for all critical future nuclear weapons functions within DoD might produce a structure wherein taking on nonnuclear responsibilities would likely be a less risky diversion.

- Even as the Executive Branch and Congress address the question of the right nuclear infrastructure for the new era, DoD should be encouraged to identify and take immediate action to reverse fragmentation, including strengthening OSD's executive authority, controls, budgetary oversight, and resource management of nuclear weapons activities throughout DoD.

RECOMMENDATIONS

The overarching recommendations of this study are:

- The United States should decide how it wants to consolidate and stabilize the overall nuclear infrastructure first (and also consider incorporating selected nonnuclear activities into that infrastructure), and then decide what to do with DNA.

- In the near term, DoD should tighten its management of nuclear matters by consolidating all of its current nuclear-weapons-related support activities (stockpile support, nuclear effects research, and threat reduction and arms control) under one senior federal executive in OSD, and within one or at most a small number of agencies reporting to that executive.

- Over the longer term, consolidation within the DoD of all U.S. nuclear-weapons-related activities should be seriously considered as a primary organizational option for a much smaller, but enduring and robust U.S. nuclear infrastructure for the 21st century.

In the near term, the DoD can significantly improve its management of nuclear matters by consolidating the executive authority for nuclear-weapons-related support (and WMD) activities (stockpile support, threat reduction and arms control, and nuclear effects research activities) under one senior federal executive in OSD. This executive should have both planning and budgetary authority for these activities.

The study also recommends that steps be initiated in FY95 to refine and validate the concepts for a consolidation of all nuclear weapon activities under the Secretary of Defense beginning in FY96. This should include a review of all DoE nuclear-weapons-related programs, as distinct from other defense programs, and a characterization of the management, operational, and technical functions, and linkages of those programs. This framework must be clear before any further steps can be considered regarding a broader consolidation of the nuclear infrastructure. The National Security Council should request that the DoD and DoE develop a program and budget description that would allow serious assessment of consolidation and scale-down options in FY96. Figure 6.1 shows our recommended schedule.

The ultimate goal is to provide for focus and leadership of the enduring nuclear-weapons-related functions. This process should begin by laying out the known problem areas and potential cost savings associated with this consolidation and characterizing the intellectual challenges of the next several decades that are better addressed within a new and streamlined national nuclear infrastructure.

Finally, we recommend a consolidation of another kind. Our review noted the number of separate DoD, DoE, OTA, GAO, and other studies dealing with aspects of the nuclear infrastructure problem. A useful first step in moving toward consolidation options would be to organize an activity bringing the study groups together to facilitate an effective interchange of viewpoints and ideas. That could be done in fall 1994.

RAND *MR442-6.1*

STEPS	FY94	FY95	FY96	FY97

• Re-engineer DNA
  – Focused/enhanced agency
  – Support infrastructure consolidation planning
		▨▨		

• Realign DoD nuclear infrastructure – Single authority for nuclear (and WMD) planning and budget decisions			▨▨	

• Review DoE national security programs
  – Nuclear activities
  – Defense activities
		▨▨▨		

• Evaluate/validate options to consolidate all DoD and DoE nuclear activities under DoD – Concepts for resolving major issues		▨▨▨		
• If analysis warrants, realign U.S. nuclear infrastructure under Secretary of Defense			▨▨▨	

Figure 6.1—Recommended Next Steps

CONGRESS AND THE DEFENSE NUCLEAR AGENCY

During the past five years, Congress has addressed itself several times to issues concerning the future size, structure, missions, and funding of the Defense Nuclear Agency. The relevant committees of jurisdiction have frequently disagreed on these matters, and the result has been a series of efforts to increase or decrease DNA's funding, to support or abolish the agency, or to provide programmatic guidance. One response to these frequent disagreements has been to call for further study of the agency's missions and future roles. The requirement for the present study is the latest manifestation of these efforts.

The Defense Nuclear Agency receives the bulk of its direct funding as an appropriation in the Research, Development, Test, and Evaluation title of the Department of Defense Appropriations bill. Other sources of funding include RDT&E verification technology, other RDT&E money in separate accounts (e.g., BMDO), operations and maintenance funds, and small amounts of procurement and military construction. In addition to direct appropriations, DNA also receives funds from other sources that DNA typically "passes through" in the form of awarded contracts. The congressional debate over the future size and structure of DNA has occurred principally in the context of the RDT&E appropriation to the agency. This appendix focuses on that account and the associated policy and programmatic direction Congress has provided.

Table A.1 illustrates the congressional appropriations for DNA's RDT&E funding (Program Element 62715H), excluding verification technology, from fiscal years 1991 through 1994. The funding profile

Table A.1

Budget Trends for the Defense Nuclear Agency, 1991–1994
(All figures are in millions of dollars)

	1991	1992	1993	1994
Budget request	355.1	441.1	409.9	288.4
HASC	355.1	378.1	359.9	259.8
SASC	385.1	416.1	372.9	288.4
Auth. conf.	326.8	367.7	372.9	238.4
HAC	326.8	341.1	362.9	238.4
SAC	326.8	290.1	382.8	0
Appropriation conference	326.8	367.7	399.7	235.0

ranges from $327 million for FY91 to a high of $373 million in FY93 before falling to $235 million in FY94.

In each year, the amounts authorized and appropriated were below the amounts requested by the administration. The increase in funding during FY92 and FY93 can be attributed to efforts to move funds from other agencies and the services to DNA, consolidating hardening and survivability technology activities formerly funded in individual services.

Beginning with the congressional funding cycle for FY92, the congressional defense committees provided direction and programmatic guidance concerning DNA's future research directions and funding priorities. These actions centered primarily on the bills authorizing appropriations, especially in the actions of the Senate Armed Services Committee (SASC). The Senate committee, in its report accompanying the FY92 Defense Authorization Bill, criticized the emphasis on strategic defense applications of certain DNA projects and reduced the requested funding by $25 million. In addition, the Senate committee strongly supported DNA work on electric armament technology, recommending an allocation of $20 million for that purpose. The House Appropriations Committee (HAC) resisted the effort to place funding for the work related to strategic defense in DNA, arguing that the funding should remain in the Strategic Defense Initiative Office (SDIO), and therefore reduced the request to $341 million. The Senate Appropriations Committee (SAC) agreed with the position of the House that testing related to strategic defenses should be funded from within SDIO, and reduced the DNA budget request still further to $290.1 million.

In the Appropriation Conference between the two houses, the authorizing committees agreed to a total authorization of $367.7 million. The conferees agreed to fund initiatives in survivability, nuclear effects, system hardness, lethality, and target hardening, initiatives that the conferees argued would be inadequately funded if not supported in the DNA budget. At the same time, the conferees criticized the DNA proposal to prepare for further underground nuclear tests (MIGHTY UNCLE and DIVINE ARCHER) and suggested delay. The amount agreed on by the authorization conference was reflected in the final appropriated amount as well.

During the FY93 funding cycle, the Congress provided even further direction to the Department of Defense concerning the programs and budget of the Defense Nuclear Agency. Again, the lead was taken by the Senate Armed Services Committee. The committee noted that the budget for DNA had increased in real terms in the years since the collapse of the Soviet Union. The committee observed that some of the increase was due to additional areas of activity being supported by DNA (e.g., treaty verification, nonproliferation), and some was because DNA was continuing to perform its traditional nuclear-related functions at the same level as during the Cold War. Although not overly critical of the agency, the Armed Services Committee suggested that a significant realignment of priorities and programs should occur within the agency, including reduced focus on strategic nuclear issues and stockpile management. The underlying reasoning on the issue of stockpile management was that a smaller stockpile should, theoretically, require fewer resources to manage.

The committee also encouraged DNA to apply its expertise to related areas of research, such as proliferation, microwave radiation, electric gun technology, and advanced conventional munitions effects.

Consistent with this point of view, the committee recommended that DNA change its name to reflect its new character and charter as an advanced weapons and munitions agency, with expertise and responsibility in the conventional as well as the nuclear arena. The committee further recommended that DNA (in its new form) report directly to the Under Secretary of Defense for Acquisition rather than to the Director of Defense Research and Engineering, and that an oversight committee within the Office of the Secretary of Defense be established. The oversight committee was further directed to conduct a comprehensive review of DNA and its missions and to report the results of that review to the Congress.

The House-Senate conference on the defense authorization bill for FY93 took note of the extensive guidance and direction provided by the Senate committee in its report but did not endorse the specific nature of the recommendations. Instead, the conferees declared that a comprehensive review of DNA's "roles, missions, funding, and management" should be conducted. The conferees ordered that the review be conducted by the Defense Science Board (DSB) and a group consisting of representatives of the Chairman of the Joint Chiefs of Staff and the Office of the Secretary of Defense (OSD). The conferees on the authorization bill declined to provide specific programmatic guidance other than a recommendation that $20 million be available for continued work on the electric gun project.

The appropriations process for FY93 was not characterized by significant activity related to DNA. The fact that the final amount appropriated exceeded the authorized amount (see Table A.1) can be attributed to a provision added to the appropriations bill in the Senate appropriating $15 million for immediate transfer to the Department of Energy. Thus, the amount actually intended for DNA paralleled the authorized amount. The Senate Appropriations Committee deleted funding for several specific projects, but indicated that the purpose of the Senate action was to hold DNA funding constant at the FY92 level in light of the "ongoing internal effort" to review the missions and resources of the agency in light of the altered international security environment.

The Defense Science Board/OSD/Joint Staff study requested by the Congress was conducted, and the results were made available in May 1993. On June 25, 1993, the Secretary of Defense transmitted the results of the study to the Congress. In a letter to the Chairman of the Committee on Appropriations, the Secretary noted that he was "satisfied with the comprehensiveness and effectiveness of the DNA program, and the manner in which it is adapting to post-cold war realities." The Secretary also noted that DNA would remain the DoD center for nuclear expertise, that it would henceforth report to the Under Secretary for Acquisition through the Assistant to the Secretary for Atomic Energy, and that a Nuclear Coordinating Committee would be established. The Secretary also stated that DNA's charter would be revised to reflect its evolving role and activities.

During the FY94 funding cycle, the authorizing committees, apparently satisfied with the directions outlined by the secretary's letter and the internal DoD report, provided funding near the requested levels. The final conference agreement reduced the requested funding (already over $100 million lower than the FY93 request) by an additional $50 million, largely in light of the decisions to suspend underground nuclear testing. The Senate Armed Services Committee, in particular, indicated its agreement with the conclusions of the internal DoD report and approved of the effort to apply DNA nuclear expertise to advanced conventional munitions.

The House Appropriations Committee followed the lead of the authorizing committees, reducing the requested amount by $50 million primarily to reflect reductions in nuclear testing costs. The Senate Appropriations Committee, on the other hand, indicated that it no longer believed the Defense Nuclear Agency could be justified as a separate agency. Instead, the committee recommended that DNA be abolished and that its essential functions be distributed among the services or other agencies. This recommendation was rejected by the Appropriations Conference, which ultimately allocated $235 million for FY94, virtually the entire authorized amount. The Conference Committee also requested that an outside review of DNA be conducted by an independent analytic organization such as RAND's National Defense Research Institute, leading directly to the present study.

BIBLIOGRAPHY

PUBLICATIONS

Agnew, Harold M., "Restructuring the U.S. Nuclear Weapons Establishment," unpublished paper, March 5, 1994.

Air Force, Military Deputy to the Assistant Secretary (Acquisition), *Report on the Defense Nuclear Agency by Direction of Goldwater-Nichols Department of Defense Reorganization Act of 1986*, June 1987.

Albright, David, et al. *World Inventory of Plutonium and Highly Enriched Uranium*, Stockholm International Peace Research Institute, Oxford University Press, 1993.

Anderson, Christopher, "Livermore Faces Forces of Change," *Science*, Vol. 264, April 15, 1994, pp. 336–338.

Aspin, Les, *Bottom-Up Review*, Washington, D.C.: U.S. Department of Defense, October 1993.

Background Information for Senate Colloquy on Defense Nuclear Agency, n.d.

Barber, Richard J., *The Advanced Research Projects Agency, 1958–1974*, December 1975.

Buchan, Glenn, *U.S. Nuclear Strategy for the Post-Cold War Era*, Santa Monica, Calif.: RAND, MR-420-RC, 1994.

Cochran, Thomas B., William M. Arkin, and Milton Hoenig, *Nuclear Weapons Databook*, Vol. 1, *U.S. Nuclear Forces and Capabilities*, Cambridge, Mass.: Ballinger, 1984.

Cochran, Thomas B., William M. Arkin, Robert S. Norris, and Milton Hoenig, *Nuclear Weapons Databook*, Vol. 2, *U.S. Nuclear Warhead Production*, Cambridge, Mass.: Ballinger, 1987.

Cochran, Thomas B., William M. Arkin, Robert S. Norris, and Milton Hoenig, *Nuclear Weapons Databook*, Vol. 3, *U.S. Nuclear Warhead Production*, Cambridge, Mass.: Ballinger, 1987.

Comments on SAC-D Report Language Regarding DNA Executive Summary (Guiding Principles), n.d.

Defense Nuclear Agency, *Catalog of Selected Programs*, Alexandria Va.: DNA, April 4, 1994.

Defense Nuclear Agency, *DNA Function Review: Status of Funds as of March 13, 1994.*

Defense Nuclear Agency, Field Command, *Field Command, Defense Nuclear Agency, Special Historical Report* [covering 1946–1985], Albuquerque, N.M.: Field Command Defense Nuclear Agency, DNA Technical Library No. DTL 070204, 1986.

Defense Nuclear Agency, Field Command, FY 94 FCDNA Joint Manpower Program, Albuquerque, N.M., n.d.

Defense Nuclear Agency, *Final Report to the Congress on Conduct of the Persian Gulf War*, Washington, D.C.: U.S. Department of Defense, April 1992.

Defense Nuclear Agency, *Radiation Test Facilities*, 3rd Ed., Washington, D.C.: U.S. Department of Defense, September 1992.

Defense Nuclear Agency, "Unclassified Listing of Official DOD Nuclear Accidents," pp. A-1 through A-4, n.d.

Defense Nuclear Facilities Safety Board, *Annual Report to Congress*, Washington, D.C., February 1993.

Defense Nuclear Facilities Safety Board, *Annual Report to Congress*, Washington, D.C., February 1994.

Defense Science Board, *Report of the Defense Science Board on Defense Nuclear Agency,* Washington, D.C.: U.S. Department of Defense, April 1993 (John M. Cornwall, Chairman).

Defense Science Board, *Review of the Defense Nuclear Agency Technology Base Program,* 17 February 1982 (John Deutch, Chairman).

Defense Science Board, Task Force on Defense Nuclear Agency Management, *Report of the Ad Hoc Defense Science Board Task Force on Defense Nuclear Agency Management,* June 1986 (Gerald W. Johnson, Chairman).

DNA Customer Requirements Process Activities (March–December 93), computer listing.

DNA Form 547, "Response to Question by Dr. Mizrahi, Defense Science Board," 16 February 1993.

DNA New Work Units FY 89-93, listing and briefing chart, provided Dr. T. Coffey, NRL, n.d.

"Domenici to Challenge Elimination of Defense Nuclear Agency," press release, October 6, 1993.

Drell, Sidney, John Foster, and Charles Townes, *Report of the Panel on Nuclear Weapons Safety,* U.S. House of Representatives, Committee on Armed Services, Washington, D.C.: GPO, HAS Report No. 15 (Drell Panel Report), December 1990.

Ernst, Col. Frederick, "Role of the Defense Nuclear Agency in the Persian Gulf Conflict," *Nuclear Survivability,* April 1992, pp. 3, 33.

Feaver, Peter D., *Guarding the Guardians: Civilian Control of Nuclear Weapons in the United States,* Ithaca and London: Cornell University Press, 1992.

Formal Requirements List, Washington, D.C.: DNA, February 5, 1994.

Foster, John S., Jr., *Statement before Military Application of Nuclear Energy Panel,* House Armed Services Committee, March 22, 1994.

Grant, John G., and Frank L. Gertcher, *Nuclear Hardening Cost Study,* RDA-TR-2-3205-4505-001, Los Angeles, Calif.: RDA LOGICON, 1993.

Healy, Melissa, "Panel to Study Closing Some of National Labs," *The Los Angeles Times*, February 3, 1994.

House Report 102-95, *Department of Defense Appropriations Bill, 1992*, Committee on Appropriations, June 6, 1991, and Senate Report 102-154, *Department of Defense Appropriation Bill, 1992*, Committee on Appropriations, September 20, 1991.

House Report 102-311, *National Defense Authorization Act for Fiscal Years 1992 and 1993*, November 13, 1991.

House Report 102-966, *National Defense Authorization Act for Fiscal Year 1993*, October 1, 1992.

House Report 103-254, *Department of Defense Appropriations Bill, 1994*, Committee on Appropriations, September 22, 1993.

House Report 103-339, *Department of Defense Appropriations Act*, November 9, 1993. Also Section 8136, Public Law 103-139, *Fiscal Year 1994 Defense Appropriations Act.*

House Report 103-357, *National Defense Authorization Act for Fiscal Year 1994*, November 10, 1993.

An Interim Report to the President by the President's Blue Ribbon Commission on Defense Management, February 28, 1986.

Jones, Anita K., *Statement to the Subcommittee on Defense Technology, Acquisition, and Industrial Base of the Senate Committee on Armed Services by Director, DR&E*, March 8, 1994.

Laboratory Emphasizing Space-Related Technology and Satellite Experiments, 24 August 1993.

Lawrence Livermore National Laboratory, *Institutional Plan*, FY94–FY99, UCAR-10076-12, December 1993.

Letter from the Honorable Les Aspin, Secretary of Defense, to Senator Robert C. Byrd, Chairman, Committee on Appropriations, June 25, 1993.

Los Alamos National Laboratory, *Institutional Plan*, FY94–FY99, LALP 93-32, December 1993.

Mahoney, Rob, *Overview—May 1993 SAGE Meeting*, Scientific Advisory Group on Effects, Defense Nuclear Agency, Draft Report, Alexandria, Va.: RDA LOGICON, 3 June 1993.

Making Contracting Work Better and Cost Less, Washington, D.C.: U.S. Department of Energy, February 1994.

Mallin, Maurice A., *The June 1993 Swedish Draft Comprehensive Nuclear Ban Treaty: Implications and Issues for Negotiations*, McLean, Va.: The Center for National Security Negotiations, SAIC, March 1994.

Martel, William C., and William T. Pendley, *Nuclear Co-Existence: Rethinking U.S. Policy to Promote Stability in an Era of Proliferation*, Air War College Studies in National Security, No. 1, April 1994.

McNamara, Robert, "Nobody Needs Nucs," *The New York Times*, February 23, 1993.

McNamara, Robert, *The Changing Nature of Global Security and Its Impact on South Asia*, address to the Indian Defense Policy Forum, published by the Washington Council on Nonproliferation, November 20, 1992.

Millot, Marc Dean, Roger Molander, and Peter A. Wilson, *"The Day After . . ." Study: Nuclear Proliferation in the Post-Cold War World, Volume I, Summary Report*, Santa Monica, Calif.: RAND, MR-266-AF, 1993a.

Millot, Marc Dean, Roger Molander, and Peter A. Wilson, *"The Day After . . ." Study: Nuclear Proliferation in the Post-Cold War World, Volume II, Main Report*, Santa Monica, Calif.: RAND, MR-253-AF, 1993b.

Millot, Marc Dean, Roger Molander, and Peter A. Wilson, *"The Day After . . ." Study: Nuclear Proliferation in the Post-Cold War World, Volume III, Exercise Materials*, Santa Monica, Calif.: RAND, MR-267-AF, 1993c.

Molander, Roger C., and Peter A. Wilson, *The Nuclear Asymptote: On Containing Nuclear Proliferation*, Santa Monica, Calif.: RAND, MR-214-CC, 1993.

Mulholland, Laura, B. Board, and D. Alderson, *Operating Charters and Missions of Selected Department of Defense Organizations*, Alexandria, Va.: DASIAC, DASIAC Special Report SR-93-018, June 24, 1993.

Narath, Albert, *Statement Before Military Application of Nuclear Energy Panel*, House Armed Services Committee, March 22, 1994.

Nitze, Paul H., "Is It Time to Junk Our Nukes? The New World Disorder Makes Them Obsolete," *The Washington Post*, January 16, 1994, pp, C1, C2.

Norris, Robert S., et al., *Nuclear Weapons Data Book*, National Defense Resources Council, 1994.

Nuckolls, John H., *Testimony Before Military Application of Nuclear Energy Panel*, House Armed Services Committee, March 22, 1994.

Nuclear Survivability, Washington, D.C.: Defense Nuclear Agency, April 1992.

Nuclear Survivability, Washington, D.C.: Defense Nuclear Agency, December 1993.

Office of Technology Assessment, U.S. Congress, *Defense Conversion: Redirecting R&D*, Washington, D.C.: GPO, 1993.

Office of Technology Assessment, U.S. Congress, *Dismantling the Bomb and Managing the Nuclear Materials*, Washington, D.C.: GPO, September 1993.

Office of the Secretary of Defense, Deputy Assistant Secretary of Defense (Administration), Directorate for Organization and Management Planning, *Reassessment of Defense Agencies and DOD Field Activities*, October 1987 (OSD Study Team) Appendix I, "Defense Nuclear Agency," pp. I-1 to I-18.

Office of the Under Secretary of Defense (Acquisition), *Program Plan for Research, Development, Test and Evaluation for Arms Control Cooperative Inspection*, 4 January 1993.

Office of the Under Secretary of Defense for Acquisition, "Report of the OSD/Joint Staff Review Group on Defense Nuclear Agency," May 1993 (includes the DSB Task Force Report).

Pelaez, Rear Admiral Marc, *Statement to the Subcommittee on Defense Technology, Acquisition, and Industrial Base of the Senate Armed Services Committee by Chief of Naval Research*, March 8, 1994.

Point Paper for Dr. Harold Smith in Preparation for Discussion with Senator Domenici, n.d.

Policy Overview, U.S. Department of Energy, n.d.

Prahalad, C. K., and Gary Hamel, "The Core Competence of the Corporation," *Harvard Business Review*, May–June 1990.

President's Blue Ribbon Commission on Defense Management, *A Quest for Excellence: Final Report to the President*, Washington, D.C., June 1986 (David Packard, Chairman).

President's Blue Ribbon Task Group on Nuclear Weapons Program Management, *Report of the President's Blue Ribbon Task Group on Nuclear Weapons Program Management*, July 1985, 2 vols. (Judge William P. Clarke, Jr., Chairman; James R. Schlesinger, Vice Chairman).

Presidential Review Directive (draft), Interagency Federal Laboratory Review, March 15, 1993.

Program Decision Package FY93, DNA Listing, March 15, 1994.

Radiation Test Facilities, Defense Nuclear Agency, Alexandria, Va.: DoD Nuclear Information Analysis Center, 3rd Ed., September 1992.

Reis, Victor H., *Testimony Before Military Application of Nuclear Energy Panel*, House Armed Services Committee, March 22, 1994.

Reis, Victor H., *Statement to the Subcommittee on Defense Technology, Acquisition, and Industrial Base of the Senate Armed Services Committee by Department of Energy*, March 8, 1994.

Research Announcement, *Upper Atmosphere Research Satellite Guest Investigator Program*, Washington, D.C.: NASA, NRA 94-MTPE-03, January 28, 1994.

Research, Development, Test and Evaluation, Defense Agencies Program Document (Fiscal Year 1995), Washington, D.C.: Department of Defense, 1 October 1993.

Rhoades, Richard, "A Different Country," lecture delivered at Los Alamos Laboratory, June 10, 1993.

Rhoades, Richard, "Atomic Logic," *Rolling Stone*, February 24, 1994.

Sandia National Laboratory, *Institutional Plan*, FY94–FY99, SAND93-2069, October 1993.

Science, Technology, and Congress: Expert Advice and the Decision-Making Process, Carnegie Commission on Science Technology and Government, February 1991.

Scott, William B., "CD-ROM Aids Analysis of Air Combat Data," *Aviation Week and Space Technology*, April 11, 1994, pp. 60–61.

Secretary of the Navy, *Report on Defense Agencies and DOD Field Activities*, August 1987, Appendix E, "The Research and Development Agencies."

Senate Report 102-113 to accompany S.1507, *Authorizing Appropriations for the Department of Defense*, Committee on Armed Services, July 19, 1991.

Senate Report 102-352, *National Defense Authorization Act for Fiscal Year 1993*, Committee on Armed Services, July 31, 1992.

Senate Report 102-408, *Department of Defense Appropriations Bill*, Committee on Appropriations, September 17, 1992. Also, see House Report 102-627, *Department of Defense Appropriations Bill, 1993*, June 29, 1992.

Senate Report 103-153, *Department of Defense Appropriations Bill, 1994*, Committee on Appropriations, October 4, 1993.

Shelton, Frank H., *Reflections of a Nuclear Weaponeer*, Colorado Springs, Colo.: Shelton Enterprises, Inc., 1988.

Sherwood, Jeff, "Alternative Futures for Department of Energy Laboratories," *DOE News Release*, March 3, 1994.

Sherwood, Jeff, "Energy Secretary Forms Task Force to Examine Future of DOE National Laboratories," *DOE News Release,* February 2, 1994.

Sigmund, Ward, *Background Information for the RAND Study of DNA,* Nuclear Weapons Council, n.d.

Singley, George T. III, *Statement to the Subcommittee on Defense Technology, Acquisition, and Industrial Base of the Senate Committee on Armed Services by Deputy Assistant Secretary of the Army for Research and Technology,* March 8, 1994.

Smith, Harold P., *Statement to the Military Application of Nuclear Energy Panel of the House Committee on Armed Services by Assistant to the Secretary of Defense (Atomic Energy),* March 22, 1994.

Statement to the Subcommittee on Defense Technology, Acquisition, and Industrial Base of the Senate Armed Services Committee by the Deputy Under Secretary of Defense for Advanced Technology, March 8, 1994.

Testimony on Air Force Science and Technology, n.d.

U.S. Congress, House, *National Defense Authorization Act for Fiscal Year 1994—Conference Report,* November 10, 1993, 103rd Congress, lst Session, Report 103-357.

U.S. Congress, House, *National Defense Authorization Act for Fiscal Year 1993—Conference Report,* July 31, 1992, 102nd Congress, 1st Session, Report 102-352.

U.S. Congress, Senate, *National Defense Appropriations Act for FY 1993,* September 17, 1992, 102nd Congress, 1st Session, Report 102-408.

U.S. Department of Defense, "Defense Nuclear Agency," Directive 5105.31, January 24, 1991.

U.S. Department of Defense, "Defense Nuclear Agency," Draft Directive, Washington, D.C., August 16, 1993.

U.S. Department of Defense, Office of the Deputy Secretary, *Report on Nonproliferation and Counteproflieration Activities*, Washington, D.C., April 1994.

U.S. Department of Energy, *Nuclear Weapons Complex Reconfiguration Study*, Washington, D.C.: Department of Energy, DOE/DP-0083, January 1991.

U.S. Department of the Army, *United States Army Nuclear and Chemical Agency: Organization and Functions*, Washington, D.C.: HQ U.S. Army, Army Reg. 10-16, 10 May 1993.

Warnke, Paul, *The Bottom-Up Review: Exaggerated Threats and Undervalued Allies*, The Center for National Security, March 1994.

Weapons Systems Lethality Project Listing, DNA PDP Summary Sheets, 2/22/94.

Wilkinson, Francis, "Power to the People," *Rolling Stone*, March 24, 1994, p. 33

1992–1993 Fact Book, Washington, D.C.: Naval Research Laboratory, NRL/PU/5240-93-238, June 1993.

BRIEFINGS

Adaptive, Lower Cost Test Facilities, Briefing Charts, n.d.

Atkins, Marvin, SAGE, January 1987, Briefing Charts, Washington, D.C.: Defense Nuclear Agency.

Bacon, David P., OMEGA Airborne Hazard Dispersal, Briefing Charts, San Diego, Calif.: Science Applications International Corporation, March 30, 1994.

Baum, Joseph, Integration of CFD/CSD Methodologies for the Simulation of Conventional Weapon Effects, Briefing Charts, McLean, Va.: SAIC, March 30, 1994.

Beers, Brian L., Radiation and Electromagnetic Effects in Systems, Briefing Charts, McLean, Va.: SAIC, March 30, 1994.

Beyster, Bob, Curt Smith, Greg Weaver, Threats Involving a Few Nuclear Weapons, Briefing Charts, January 31, 1994.

Bisson, Arthur E., Navy Programs, Briefing Charts, Washington, D.C.: Office of Naval Research, November 19, 1993.

Briefing Charts, DUSD(AT)SASC, March 8, 1994.

Briefing to Industry, Phillips Laboratory, August 13, 1991.

Dassler, Col. Bill, and Capt. Pete Selde, Assessments and Training, Briefing Charts, Field Command Defense Nuclear Agency, September 27, 1993.

DECADE Simulator, Briefing Charts, n.d.

Defense Nuclear Agency, "1994 Program Strategy Review," Briefing Charts, 1994.

Defense Nuclear Agency, Cooperative Threat Reduction Program, Briefing Charts, January 1994.

Defense Nuclear Agency, Corporate Board Review of DNA Response to Customers, Briefing Charts, April 5, 1994.

Defense Nuclear Agency, Directorate for Information Management, Briefing Charts, December 2, 1993.

Defense Nuclear Agency, Directorate for Test Simulator Information, Briefing Charts, n.d.

Defense Nuclear Agency, Field Command, INWS Mission Statement, Briefing Charts, n.d.

Defense Nuclear Agency, Field Command, RAND Corporation In-brief, March 1994.

Defense Nuclear Agency, Field Command, RAND Corporation Background Data, Albuquerque, N.M.: Briefing Charts, March 1994.

Defense Nuclear Agency, JUPITER, Briefing Charts, n.d.

Defense Nuclear Agency, Overview Briefing, Briefing Charts, n.d.

Defense Nuclear Agency, Overview Briefing, Fiscal Year 1995 Budget Presentation, March 1994.

Defense Nuclear Agency, Radiation Directorate, Briefing, March 1994.

Defense Nuclear Agency, Safety Assessment Division, Briefing Charts, undated.

DNA's Top Ten Contractors, FY89–FY93, Briefing Charts, n.d.

Federal New Services, Aspin Nuclear Review, Press Briefing, October 29, 1993.

FY 1995 Congressional Budget Request, Press Briefing Charts, U.S. Department of Energy, February 1994.

Good, Earl R., Overview: Phillips Laboratory, Briefing Charts, September 1993.

Hagemann, Maj. Gen. Kenneth, "Course of Change," Defense Nuclear Agency, February 7, 1994.

Herman, Major David, USAF, "Preserving Air Force Expertise," Briefing, U.S. Air Force Air Materiel Command, Operations Support Branch, NW 10, Joint Advisory Committee, 12 January 1994.

Impact of Abolishing DNA, Briefing Charts, n.d.

Latko, Robert, DNA Support to Theater Nuclear Planning, RAND-SAIC Technical Interchange Meeting, Briefing Charts, San Diego, Calif: SAIC, 30 March 1994.

Lawson, Col. Harlan A., Briefing to the RAND Review Panel, Briefing Charts, March 25, 1994.

Lippincott, Capt. L. H. (USN), FCDNA Posture Review Briefings, Briefing Charts, May 1994.

Mangan, Dennis L., Sandia National Laboratories: On-Site Monitoring Program, Briefing Charts, n.d.

Nuckolls, John H., Briefing to Military Application of Nuclear Energy Panel, House Armed Services Committee, March 22, 1994.

On-Site Inspection Agency, Command Briefing, Briefing Charts, n.d.

Operations Directorate, Briefing Charts, n.d.

Option #2, DNA Functions Performance by Services & DoD Labs, Briefing Charts, n.d.

OSD/Joint Staff and Defense Science Board Task Force Defense Nuclear Agency Review, Executive Briefing, May 1993.

Phillips, Garry T., Nuclear Weapons Effects: Shock Physics, RAND-SAIC Technical Interchange, Briefing Charts, San Diego, Calif: SAIC, March 30, 1994.

Powell, James E., Nuclear Weapon Effects Testing Discussion for RAND Study of DNA, Briefing Charts, March 24, 1994.

Powell, James E., Nuclear Weapon Effects Testing Discussion for RAND Study of DNA, Briefing Charts, Sandia National Laboratories, March 24, 1994.

RAND, A Top Down Assessment for the Management of the U.S. Nuclear Infrastructure, Briefing Charts, April 8, 1994.

RAND, Defense Nuclear Agency (DNA) Functions, An Assessment of Future Options, Briefing Charts, presented to Harold Smith, Santa Monica, Calif., March 31, 1994.

RAND, Review of the Defense Nuclear Agency, Task 1: Future Environments, Briefing Charts, February 7, 1994.

RAND, Task 2: Organizational Assessment, Briefing Charts, 31 March 1994.

RAND, Task 3: Synthesis and Analysis, Briefing Charts, n.d.

Richlin, Maj. Debra, and Robert C. Webb, Defense Acquisition Board (DAB) Support, Briefing Charts, n.d.

Sandia National Laboratories, On-Site Monitoring Program, Briefing Charts, n.d.

Sandia National Laboratories, Operations Center—Dual Mission, Briefing Charts, January 24, 1994.

Sandia National Laboratories, Overview, Briefing Charts, February 4, 1994.

Sandia National Laboratories, Surety Assessment Center Interactions with DNA, Briefing Charts, March 23, 1994.

Sandia National Laboratory, DNA/Sandia Surety Program, Briefing Charts, n.d.

Shirley, Clinton G., Nuclear Surety, Briefing Charts, March 23, 1994.

U.S. Department of Defense, Combined Battlefield Environmental Effects: Executive Steering Committee Kickoff Meeting, Briefing Charts, April 20, 1994.

Wilson, Howard L., RAND Energy Coupling Review, Briefing Charts, San Diego, Calif: SAIC, March 30, 1994.

Woolson, William, CORES (Radiation Effects), RAND-SAIC Technical Interchange, Briefing Charts, San Diego, Calif: SAIC, March 30, 1994.

LETTERS AND MEMORANDUMS

Arnold, Timothy, FAX to Maurice Mizrahi, February 24, 1993.

Aspin, Les, letter to Senator Sam Nunn, June 25, 1993.

Atkins, Marvin, letter to Maurice Eisenstein, April 15, 1994.

Bracken, Paul, memorandum to Elwyn Harris and DNA Group, April 5, 1994.

Butler, George L., letter to Representative John P. Murtha, 7 October 1993.

Carpenter, H. Jerry, letter to Rich Mesic, dated April 8, 1994.

Coyle, Phil, FAX to Bryan Gabbard/Bob LeLevier, February 25, 1994.

Coyle, Phil, memorandum to Bryan Gabbard/Bob LeLevier, Subject: Consolidation, March 21, 1994.

"Defense Nuclear Agency Legislative Strategy," memorandum for Dr. Harold Smith, October 6, 1993.

Diaz, Angela Phillips, "Draft NSTC Committee Taxonomies," memorandum to Distribution, February 15, 1994.

Downie, Col. Mike, FAX to Maury Eisenstein, March 29, 1994.

Eisenstein, Maurice, "Reorganization of U.S. Nuclear Weapons Program and DNA's Future," memorandum to Bryan Gabbard, Rich Mesic, Elwyn Harris, and John Friel, February 21, 1994.

Foley, Kate, U.S. Department of Energy, Defense Programs, "Historical Funding for Defense Programs," with excerpts, "History of Defense Program Funding, FY1980–FY1995," April 1994.

Forster, LTG William H., "Army Position on Status of the Defense Nuclear Agency," memorandum for Assistant to the Secretary of Defense for Atomic Energy, 12 October 1993.

Gilinsky, Victor, "Consolidating DoD/DoE Nuclear Function," Memorandum to Bryan Gabbard, Phil Coyle, Bob LeLevier, 10 April 1994.

Gritton, Eugene C., "Meeting with Gary Denman, Director, ARPA, Paul Kozemchak and Randy Greesang, Special Assistants to the Director, ARPA, November 3, 1993," memorandum for the record, November 5, 1993.

Hagemann, Kenneth, "Congressional Action Affecting DNA," memorandum to all directorates/separate offices, October 1, 1993.

Harris, Elwyn, "Background on DNA Study: Organizational Assessment," memorandum to Paul Bracken, March 24, 1994.

Harris, Elwyn, "Contractor List from Major John Keyma, DNA," memorandum to List, March 14, 1994.

Harris, Elwyn, letter to Mike Dove, February 28, 1994.

Harris, William R., "Institutional Reforms to U.S. Nuclear Weapons Training Safety, and Security . . ." memorandum, April 4, 1994.

Harris, William R., "International Functions of the Defense Nuclear Agency," memorandum, March 1994.

Inouye, Daniel, and Ted Stevens, letter to William Perry, October 4, 1993.

Jones, Dan, "Statistical Analysis of Trends in Military Manpower," memorandum to Elwyn Harris, John Friel, Maurice Eisenstein, April 15, 1994.

Jones, Dan, "Trends in Armed Services 'Nuclear Personnel,'" memorandum to Elwyn Harris, John Friel, Maurice Eisenstein, April 15, 1994.

Jones, Dan, "Trends in the US Nuclear Stockpile," memorandum to Elwyn Harris, John Friel, Maurice Eisenstein, April 18, 1994.

Laird, Melvin, and Harold Brown, letter to Daniel Inouye, October 6, 1993.

Layson, Bill, FAX to Maury Eisenstein, March 24, 1994.

Lippencott, L. H., Capt., letter to William R. Harris, Subject: Summary of Field Command DNA History, April 20, 1994.

Mesic, Richard, "PACCOM," memorandum to Bryan Gabbard, April 4, 1994.

Nuclear Weapons Council, "Plan for a Stockpile Stewardship Under a Test Ban," memorandum, n.d.

Office of Science and Technology Policy (OSTP), "Presidential Review Directive on an Interagency Review of Federal Laboratories," memorandum, May 5, 1994.

Peltason, J. W., letter to The Honorable Hazel R. O'Leary, March 2, 1994.

Perry, William J., letter to Senator Ted Stevens, 11 October 1993.

Reis, Vic, and Jack Keliher, "Draft Strategic Plan," memorandum to National Security Strategic Planning Team, November 30, 1993.

Scarborough, Capt. O. D., III, "Potential Dismantlement of the Defense Nuclear Agency," memorandum for Principal Deputy Assistant to the Secretary of Defense (Atomic Energy), 1 October 1993.

Sigmund, Ward, "Background Information for the RAND Study of DNA," memorandum, n.d.

Smith, Harold P., Jr., letter to David Gompert, 28 February 1994.

Solomon, Kenneth, "DNA Study," memorandum to Elwyn Harris, John Friel, and Richard Mesic, February 18, 1994.

Solomon, Kenneth, "DNA Study: Revisited Draft February 21, 1994," memorandum to Elwyn Harris, John Friel, and Richard Mesic, February 21, 1994.

Solomon, Kenneth, "DNA Survey—Purpose, Instructions, Confidentiality," memorandum to B. Gabbard, E. Harris, S. Hosek, J. Friel, and R. Mesic, February 18, 1994.

Solomon, Kenneth, "Pilot Survey for DNA Project," memorandum to DTP staff, February 23, 1994.

Toomay, J. C., "DNA Study—Comments on DNA Functions," memorandum for Bob LeLevier, March 8, 1994.

Yockey, Donald, "Soviet Nuclear Threat Reduction Act of 1991," memorandum to list, 31 March 1992.

INTERVIEWS

Allen, General Lew (USAF, ret.), Project Senior Advisor, February 25, 1994.

Andresen, Steve, Director for Defense Policy and Arms Control, National Security Council, April 13, 1994.

Bell, Robert G., Special Assistant to the President, Senior Director for Defense Policy and Arms Control, National Security Council, March 23, 1994.

Beyster, Robert, Chief Executive Officer, Science Applications International Corporation, March 30, 1994.

Birely, John, Deputy for Cooperative Threat Reduction, Assistant to the Secretary of Defense (Atomic Energy), March 3, 1994.

Brewer, Colonel Ben A., Director, Office of Intelligence and Security, Defense Nuclear Agency, March 9, 1994.

Brode, Hal, Vice President, Pacific Sierra Research, March 22, 1994.

Celec, Fred, Deputy Director, Operations Directorate, Defense Nuclear Agency, February 16, 1994, March 9, 1994.

Cliff, Gene, Office of the Assistant to the Secretary of Defense (Atomic Energy), March 3, 1994.

Coffee, Tim, Deputy Director, Naval Research Laboratory, April 12, 1994.

Cole, Brigadier General J. L., Nuclear Weapons Council, Standing Committee, April 12, 1994.

Crawford, Lieutenant Colonel Mike, Office of the Assistant to the Secretary of Defense (Atomic Energy), April 12, 1994.

Davidson, Charles, Director, U.S. Army Nuclear and Chemical Agency, March 11, 1994.

Denman, Gary, Director, Advanced Research Projects Agency, February 18, 1994.

Deutch, John M., Deputy Secretary of Defense, February 15, 1994.

Downie, Colonel Michael, Deputy Director, Technology Applications Directorate, Defense Nuclear Agency, February 7, 1994, March 24, 1994.

Drake, Jim, R and D Associates, March 22, 1994.

Drennenan, Colonel Jerry M., Assistant Director for Nuclear Operations, Defense Nuclear Agency, March 9, 1994.

Dur, Rear Admiral Phillip, Nuclear Weapons Council, Standing Committee, April 12, 1994.

Evans, Mary, Deputy Director, Conventional Arms Control and Compliance, March 2, 1994.

Fahey, Andy, Radiation Division, Defense Nuclear Agency, February 16, 1994.

Foord, Robert, Central Intelligence Agency, March 2, 1994.

Ford, Colonel Grover, U.S. Army Nuclear and Chemical Agency, March 11, 1994.

Ford, Major General Phillip, J4, STRATCOM, March 14, 1994.

Foster, John, Former Chairman, Defense Science Board, April 19, 1994.

Freedman, Jerry, Deputy Assistant for Nuclear Matters, Office of the Assistant to the Secretary of Defense (Atomic Energy), March 3, 1994.

Fuertes, Major Louis L., Chief, Nuclear Accident Branch, Operations Directorate, Defense Nuclear Agency, March 9, 1994.

Gershwin, Larry, National Intelligence Officer for Science and Technology, National Intelligence Council, March 9, 1994.

Godsey, Brigadier General Orin, J36, STRATCOM, March 14, 1994.

Goldberg, Captain Marc, Executive Assistant, Office of the Assistant to the Secretary of Defense (Atomic Energy), April 12, 1994.

Hagemann, Major General Kenneth L., Director, Defense Nuclear Agency, February 7, 1994, April 12, 1994.

Hecker, Sid, Director, Los Alamos National Laboratory, March 14, 1994.

Hoehn, William, Professional Staff Member, Senate Armed Services Committee, February 4, 1994, April 1, 1994.

Horton, Barry, Principal Deputy Assistant Secretary of Defense (Command, Control, Communications and Intelligence), April 22, 1994.

Ioffredo, Mike L., Deputy Director for Strategic and Space Programs, Program Analysis and Evaluation, February 18, 1994, March 3, 1994.

Keating, Ken, On-Site Inspection Agency, March 11, 1994.

Kelly, Stewart, Information Management Directorate, Defense Nuclear Agency, March 9, 1994.

Kingman, Colonel Rick, Defense Nuclear Agency Liaison, STRATCOM, March 9, 1994.

Kirby, Andy, Central Intelligence Agency, March 2, 1994.

Leclaire, Dave, Planning and Resources, Department of Energy, March 11, 1994.

Lees, Fred, Defense Intelligence Agency, March 4, 1994.

Lennon, Peter, Professional Staff Member, Senate Appropriations Committee, February 4, 1994, April 1, 1994.

Linger, Don, Director, Test Directorate, Defense Nuclear Agency, February 16, 1994.

Linhard, Major General Robert, J5, STRATCOM, March 14, 1994.

Lyle, Gregory K., Nuclear Operations Directorate, Defense Nuclear Agency, March 9, 1994.

Mansfield, John E., Professional Staff Member, Senate Armed Services Committee, February 4, 1994, April 1, 1994.

McFarland, Cliff, Director, Shock Physics Directorate, Defense Nuclear Agency, February 16, 1994.

McMillan, Michael H., On-Site Inspection Agency, March 11, 1994.

Meese, Admiral Richard, J52, STRATCOM, March 14, 1994, April 12, 1994.

Mehan, Colonel Ed, Advanced Research Projects Agency, March 10, 1994.

Miller, Franklin C., Senior Counselor to the Assistant Secretary of Defense (International Security Policy), Under Secretary of Defense for Policy, March 3, 1994.

Minichiello, Lee P., Deputy Director, Strategic Arms Control and Compliance, Strategic and Space Systems, Under Secretary of Defense (Acquisition and Technology), March 2, 1994.

Mullady, Colonel Brian P., Deputy Director, On-Site Inspection Agency, March 11, 1994.

Nelson, Brigadier General H.W., Nuclear Weapons Council, Standing Committee, April 12, 1994.

Norris, Captain W.L., Chief of Nuclear Division, J5, Office of the Joint Staff, March 4, 1994.

Perett, Robert, Lawrence Livermore National Laboratory, April 11, 1994.

Preston, Captain Randy, J5, Office of Joint Staff, March 4, 1994.

Reynolds, Colonel R.V., AQQ, Office of the Secretary of the Air Force, April 12, 1994.

Richlin, Major Debra L., Technology Applications Director, Defense Nuclear Agency, March 9, 1994.

Reis, Victor, Assistant Secretary of Energy, Defense Program, March 11, 1994, March 14, 1994.

Scellero, Fred, Central Intelligence Agency, March 2, 1994.

Schneiter, George R., Director, Strategic and Space Systems, Office of the Under Secretary of Defense for Acquisition and Technology, March 4, 1994.

Shore, Michael, Director, Technology Applications Directorate, Defense Nuclear Agency, February 16, 1994, March 9, 1994.

Shuler, William, Deputy for Counter-Proliferation, Office of the Assistant to the Secretary of Defense (Atomic Energy), March 3, 1994.

Sigmund, Ward, DoE Representative, Office of the Assistant to the Secretary of Defense (Atomic Energy), March 3, 1994.

Slocombe, Walt, Principal Deputy Under Secretary of Defense for Policy, March 3, 1994.

Smith, Harold P., Assistant to the Secretary of Defense (Atomic Energy), March 31, 1994, April 12, 1994, April 31, 1994.

Soper, Gordon K., Principal Deputy, Office of the Assistant to the Secretary of Defense (Atomic Energy), March 3, 1994, April 12, 1994.

Tolin, Brigadier General Anthony, Nuclear Weapons Council, Standing Committee, April 12, 1994.

Tompkins, Lieutenant Colonel Dan T., Emergency Action Branch, Operations Directorate, Defense Nuclear Agency, March 9, 1994.

Tucker, Colonel Randy, On-Site Inspection Agency, March 11, 1994.

Ullrich, George, Deputy Director, Defense Nuclear Agency, February 7, 1994, March 24, 1994.

Webb, R.C., Chief, Electronics and Systems Technology Division, Radiation Sciences Directorate, Defense Nuclear Agency, March 9, 1994.

Wheelon, Albert D., Project Senior Advisor, February 23, 1994.

Woodruff, Larry, Lawrence Livermore National Laboratory, March 28, 1994.

York, Herb, Project Senior Advisor, February 23, 1994.